D1860521

MANUAL

OF

AGRICULTURE FOR INDIA

Owing to the antiquity and rarity of this book, certain pages are regrettably absent, and readability is compromised due to the blurring of the original text. If you possess any insights regarding the availability of these missing sections, kindly inform us. Your assistance is crucial in ensuring the completion of this significant work for the benefit of future generations.

MANUAL
OF
AGRICULTURE FOR INDIA

FREDERICK POGSON

GYAN

Published by

Gyan Publishing House
5, Ansari Road
Daryaganj, New Delhi-110002
Phone: 011-47034999, 9811692060
E-mail: books@gyanbooks.com

Distribution Network
gyanbooks.com
India, USA, Canada, UK, Australia

© Publisher

All rights reserved. No part of this work may be reproduced, stored, adapted, or transmitted in any form or by any means, electronic, mechanical, photocopying, micro-filming, recording or otherwise, or translated in any language, without the prior written permission of the copyright owner and the publisher.

ISBN: 978-93-7133-618-5 (PB)

First Published, 1883

2nd Impression 2025

Printed at: Gyan Press, Delhi.

The book is sold subject to the condition that it shall not, by way of trade or otherwise, be lent, resold, hired out, or otherwise circulated without the prior publisher's written consent.

MANUAL OF AGRICULTURE FOR INDIA
Author: FREDERICK POGSON

MANUAL

OF

AGRICULTURE FOR INDIA.

BY

LIEUTENANT FREDERICK POGSON
(*Her Majesty's Bengal Army*),

HONORARY MEMBER AGRI-HORTICULTURAL SOCIETY OF INDIA,
AND AGRI-HORTICULTURAL SOCIETY OF BIJNOR, NORTH WESTERN PROVINCES.
AUTHOR OF "POGSON'S INDIAN GARDENING," AND OF "THE ANCIENT
WEIGHTS AND MEASURES OF ISRAEL IDENTIFIED WITH THOSE
OF PREHISTORIC AND HISTORIC ENGLAND."

CALCUTTA:
THACKER, SPINK AND CO.
BOMBAY: THACKER & CO., LIMITED. MADRAS: HIGGINBOTHAM & CO.
LONDON: W. THACKER & CO.
1883.

PREFACE.

IT has been officially announced that the population of India is steadily increasing, whilst the productive powers of the long-neglected and over-worked soil are as steadily decreasing.

The writer has long anticipated such a result. The very purity of the water of the canals, used for irrigation, has gradually dissolved fertilizing substances which, held in aqueous solution, have descended into the subsoil beyond the reach of the roots of growing crops, and these have biennially carried off a sensible quantity of fertilizers, producing more or less exhaustion of the soil, as indicated by the reduced yield of grain of all kinds per acre. The most notable instance of this soil exhaustion has just been announced in public prints, and it is but the beginning of the end, if matters agricultural are neglected. In the *Delhi Gazette* of the 30th January, 1883, we are told that "The opium crop has almost entirely failed throughout a great portion of the Chupra and Tirhoot district." This failure means a loss of at least one million sterling of State income, which would not have taken place had the Opium Department, when providing wells for the poppy cultivators, placed it in their power to make and use the "POPPY MANURE" mentioned in this work, the formula having been published in January, 1870, and being within the reach of all concerned.

The flooding of fertile lands with canal water has in numerous districts brought saline matters from the subsoil to the surface, producing sterilization. To meet this evil all kinds of empirical experiments have been tried, all terminating in the completest of failures. The true and only remedy, however, has never been tried, and unless the powers of the Excise Department are curtailed, may remain untried. The remedy is simple and practical, and is as follows, viz., All lands sterilized by *Kuller* should, after being ploughed, be sown with the seed of the *Lana* (*Salsola soda*) of the Punjab. The land so sown should not be irrigated, and the seed should be sown before the rains commence. The crop will remove a considerable quantity of salt and sulphate and carbonate of soda from the soil. The cultivation to be carried on annually (all irrigation being prohibited), until the land refuses to grow any more *Lana*, owing to exhaustion of the *Kuller*. The reclaimed land should now be sown with turnips, which will remove much sulphuric acid, and next autumn the soil will have been sufficiently sweetened to grow a crop of wheat.

To guard against failure the land should be limed with *Kunkur* dust previous to being sown with wheat, and the system of over-irrigation should be most carefully guarded against.

The system of reclamation indicated may be carried out by the zemindars if supplied with *Lana* seed, and the Excise Department should allow them to use the green leaves as cattle fodder.

The writer has foreseen that which is now coming to pass as regards the population. The summary suppression of female infanticide has borne fruit; and unless the increase of population is met with an increased supply of food from the same extent of land, which under a bad system of husbandry sufficed for a smaller population, the new danger to be encountered cannot be successfully met and disposed of.

The zemindars and ryots throughout the length and breadth of the land are now about to be called upon "*to make two blades of grass grow where one only grew before,*" and, to enable them to do so, practical instruction in agriculture is essentially necessary. The present Manual places ample instructive information at the disposal of all who can read and understand English, irrespective of rank, position, caste, creed, or colour. But the zemindars and ryots of India, who can neither read nor write their mother tongue, and are too old to learn, may be counted by the million; and for the benefit of these classes and their offspring, the rising generation, an abstract Hindee translation of this Manual will in due course be published if the project meets with the support of the Indian Government—to be followed by a translation therefrom in Urdu and Punjabee.

J. FRED. POGSON.

DEHRA DOON,
February 3, 1883.

CONTENTS.

MANUAL OF AGRICULTURE FOR INDIA.

INTRODUCTORY.

1. *What is Agriculture?*

In Hindoostan, Agriculture is the occupation of many millions of men who are called zemindars and ryots, and it is the art of cultivating the soil to the best advantage, so as to enable it to produce the finest and largest crops of every description at the lowest cost, and without exhausting and impoverishing the soil.

2. *What ought the zemindar especially to know in order to attain this end?*

The zemindar should know how to plough his land properly, and to determine its quality. He should also know how to make and apply manures to the crops he raises, and to do this properly he should have a practical knowledge of the nature and composition of the various crops which may be grown in Hindoostan.

3. *How is Agriculture practised in Hindoostan?*

In the most primitive and least remunerative form, the poverty, superstition, and ignorance of the people causing numbers of them to look upon a plentiful harvest as a forerunner of some approaching domestic calamity or death, only to be averted by giving away all grain and produce above the ordinary maximum.

It is also feared that the regular production of heavy crops would lead to an enhancement of the rent, or revenue paid to the State; whilst in practice the person chiefly and directly benefited by an abundant harvest is the village usurer and bunneeah. Hence any improvement in Agriculture is avoided, and the agricultural classes remain ignorant and impoverished.

4. *How is the condition of the ryot and zemindar to be ameliorated, and his ignorance, apprehensions, and poverty removed?*

By giving him practical instruction in Agriculture, and placing it in his power to borrow money at a reasonable rate of interest; say from six to twelve per cent. simple interest, instead of 37½ per cent. compound interest, the present minimum rate charged by the bunneeah, or village usurer.

5. *Who is to provide the money?*

Agricultural banks, to be formed by titled and untitled Talookdars and zemindars, who would be associated with European and native bankers, the Government supporting their efforts and joining the undertaking as a shareholder, or contributor of cash deposits (with eighteen months' notice of withdrawal), bearing interest at five per cent. Such banks should also be allowed to lend money on the security of gold and silver jewellery.

6. *How is agricultural instruction to be imparted to men who can neither read or write, but who have acquired from proverbs, oral tradition, and hereditary verbal in-*

struction, such knowledge as they do possess on agricultural matters?

By translating one or more practical works on Agriculture from English into Hindee and Urdu, and causing them to be read by educated men to the adult village population, assembled once a week for the purpose; and, in addition, by placing such translations in the keeping of village authorities for use as books of reference and instruction, as may be necessary. By this plan the village *Putwaree*, the village teacher or *Pundit*, the village bunneeah, or any one who could read Hindee or Urdu, would be able to seek out the information desired, and thus instruct and enlighten the applicant be he ever so illiterate.

CHAPTER I.

ORIGIN AND GENERAL CHARACTER OF SOILS.

7. *Of what two principal parts do soils consist?*

All cultivable soils consist of an organic part, which if placed in the fire will burn away; and an inorganic, or mineral part, which will not burn.

8. *Whence is the organic part of the soil derived?*

It is derived from the remains of animals, insects, and minute visible and invisible organisms of various kinds; from the dung of animals, birds, caterpillars, and worms, and from the roots, leaves, and stems of decayed weeds and plants.

9. *What constitutes the mineral, or inorganic part of soils?*

1. Sand; 2. Clay; 3. Lime; 4. Magnesia; 5. Oxides of Iron; 6. Oxides of Manganese; 7. Potash; 8. Soda; 9. Sulphuric acid; 10. Phosphoric acid; 11. Carbonic acid.

10. *How many kinds of soils are there in Hindoostan, or India?*

Officially five kinds are recognized, three of which are represented by their quality, or the numbers 1, 2, and 3, barren, and waste. But properly speaking, India possesses a greater variety of soils than any country but China, for it has tea, cotton, sugar, spice, and tobacco soils, in addition to the soils met with in Europe.

11. *Describe the soils indicated by comparison of the numbers,* 1, 2, *and* 3, *giving their common European names.*

Wheat, barley, maize, sorghum, gram, and other grain-producing lands named after such crops, come under the first, as also do cotton, oil seeds, and sugar-cane land. "Light soils," such as produce inferior grain and pulse crops, under the second; and "sandy soils" under the third. There are also very rich soils, fertile soils which have become sterile owing to saline impregnation, and absolutely barren soils.

12. *What is a rich soil?*

A rich soil generally contains five per cent., or one-twentieth its weight, of organic matters, in combination with a due proportion of clay, sand, oxide of iron, lime, magnesia, and certain other fertilizing substances.

13. *What is a fertile soil?*

In the plains of India all first-class (No. 1) soils, which continuously grow fair crops without manure, are naturally fertile soils; second and third class soils are those which must be manured to remain fertile, and their productiveness will depend on the portion of mineral fertilizers present in each description of soil.

14. *What is a light soil?*

Light soils are such as contain a large proportion of of sand, with little clay, and other mineral substances in proportion; organic matters are present in moderate quantity, and manuring is, as a rule, necessary.

15. *What is a pure (cultivable) sandy soil?*

The sandbanks, formed in the beds of the Ganges, Jumna, Sutledge, and other Indian rivers, constitute pure sandy soils, and in certain localities they are turned to very profitable account by being cultivated with melons, water melons, and some vegetables suited to the soil. All these crops are removed before the water in the river rises and covers the sandbanks, which are submerged, and so remain till the rains are over in the hills, when the water subsides and they reappear.

If artificial irrigation be available, these crops may be successfully grown on inland pure sandy soils.

16. *What is a sugar-cane soil?*

The black cotton soil of the Deccan, and similar calcareous soils, wherever existing, may, on being properly manured, be exalted to the position of natural sugar-cane soils, which is a brick loam (as in Jamaica and the West Indies), extremely rich in lime and vegetable or organic matter, and contains very little salt, the presence of which is as injurious to sugar-cane as it is to the sugar-beet of Europe. It is for this reason that human manure does not suit sugar-cane, as it contains more salt than other descriptions of manure.

17. *What is a tobacco soil?*

All soils which are extremely rich in iron, lime, and and organic matters, and are of a red, reddish-brown, and liver colour, may be called natural tobacco soils. They will also grow superior opium, *i.e.*, poppy crops. But without proper manuring the best soil will not produce superior tobacco, because the soil cannot supply the very large proportion of ammonia and potash needed by this plant. Human manure suitably enriches tobacco soils, and benefits the growing tobacco crops.

18. *What is a naturally barren soil?*

The absolutely barren soils of India are, as a rule, called "*Oosur*" by the agriculturists, and being very often hard indurated clay soil, neither weed or plant will grow on them.

19. *What is a sterile soil?*

All lands which from natural causes are highly impregnated with saline matters are sterile as far as grain, root, and other crops are concerned but as such soils will produce the *Salsola soda* plant (the valuable *Lana* of the Punjab), the *Salicornia fruticosa*, and also the edible *Salsola Indica*, they cannot be called sterile. The first and second yield barilla, or soda ash, when burned, and the third, according to Professor O'Shaughnessy, produces an abundant crop of "green leaves, universally eaten by all classes of natives who live near the sea, and are reckoned very wholesome. The leaves of this plant alone saved many thousand lives during the famine of 1791–2–3."—*O'Shaughnessy.*

The *Salsola tragus*, which yields an ash containing from 25 to 40 per cent. of carbonate of soda, would also grow luxuriantly on the so-called sterile saline lands of India, and which in the Punjab cover hundreds of square miles of land. Spain is enriched by its barilla trade. India has none!

20. *What is meant by a sterilized soil?*

In various parts of Upper and Northern India (Punjab) the surface soil was very fertile until canal irrigation was introduced. The water reached the subsoil, which contained large quantities of saline matters, and as the ground dried, these rose to the surface by capillary attraction, and rendered the soil unfit for the growth of all field crops; consequently the once fertile land became sterilized, and so continues.

The sterilizing salt which is a compound of common salt, sulphate of soda, and carbonate of soda, is called *Rhae* in Oudh and the North-Western Provinces, and *Kuller* in the Punjab. There is another saline efflorescence met with in wheat and barley fields during the winter which is not a sterilizer; it is carbonate of soda mixed with nitre, or crude nitrate of soda. It is common at Shamleh, near Meerutt, North-West Provinces.

21. *What is a nitrous soil?*

All lands which from natural causes have become impregnated with common salt and saltpetre, or nitre, are called nitrous soils. The *Salsola soda* plant will not grow on such soils, but if ploughed and sown with the seed of the saltpetre plant (*Reaumuria vermiculata*), they will yield a very valuable crop which may be sun-dried, reduced to powder, and so used as a mineral manure, or be operated upon for the production of nitrate of potash and salt.

22. *What are lake, or subaqueous soils?*

The beds of Indian lakes and *Jheels* possess soils remarkable for their richness in fertilizing matters; so much so that fields and lands in their vicinity are often manured by the application of such soil. The richness is caused by this soil being the grave of countless millions of aquatic *Infusoria*, of young and small fish, and frogs, and the final resting-place of the vast hordes of microscopic organic life whose world is water.

In such subaqueous soils, the *Singhara*, or water-nut (*Trapa bispinosa*), grows to perfection. This nut, like the potato, is an annual, and once planted will reproduce itself. It may also be regularly cultivated by the seed nuts being planted in the beds of depressions after the rain has filled them with water.

Acting on the writer's suggestion made some years ago, this valuable food nut has been successfully introduced into Australia, and similar success has been attained in America, where the seed nuts are raised and offered for sale by seedsmen.

The lakes of Cashmere, 5,000 feet above the level of the sea, annually supply a crop of water nuts sufficient to feed 30,000 people for five months and yield a revenue of over £10,000 a year.

The utilization of all subaqueous soils (if under Government control) in the manner indicated would be productive of much good, and especially so in the Madras Presidency, so remarkable for its immense freshwater lakes and peculiar liability to drought, the precursor of scarcity and famine.

23. *In Question 9, the mineral or inorganic parts of soils have been stated: more information is desired, therefore explain: What is lime?*

When limestone, including *Kunkur*,[1] is burned, its contained carbonic acid gas is expelled by the action of the fire, and that which remains is called lime, or quicklime, or in Hindee, *Guttee choona*, and in Urdu, *Ghair*, *Boojhya*, *Hooah choona*.

Fifty seers of limestone of the best quality, before being burned into lime, contain 28 seers of lime and 22 seers of carbonic acid gas. This gas is latent in all limestones.

24. *What is slaked lime?*

When water is poured upon quicklime it is rapidly

[1] *Anglicè*, Nodular limestone.

absorbed, and in a little while the lime gets hot, begins to swell, steams, and gets as hot as fire, and then commences to fall in pieces. These presently fall into a fine white powder, which when cold is called slaked lime, and in Hindee, *Choona*, in Urdu, *Bazaree choona.*

25. *What weight of quicklime is obtained from a ton of limestone?*

A ton of pure limestone weighs exactly 27 maunds, 8 seers, 14 chittacks, and 200 grains; and when properly burned it will yield of quicklime 11¼ cwt., or 15 maunds, 12 seers, and 8 chittacks.

26. *Does quicklime increase in weight when slaked?*

Yes; one ton of pure quicklime becomes 26⅓ cwt. of slaked lime, or 35 maunds, 33 seers, 11 chittacks, and 233⅓ grains. Hence the increase of weight is equal to 8 maunds, 24 seers, 13 chittacks, and 33⅓ grains; and as nearly the whole of this increase of weight is due to the water absorbed by the quicklime, the agriculturist who purchases lime should stipulate to be supplied with unslaked lime, and slake it on his own premises without delay, and when cool store it under shelter.

27. *Will all kinds of sea and fresh water shells, as well as all kinds of coral, yield quicklime if burned?*

Yes; they will yield, when burned, very superior, pure quicklime, and they are much more easily fired and converted into lime, owing to the thinness of the shells and coral branches, as compared with limestone and *Kunkur.*

28. *Why should lime be applied to soils which, from constant cropping, have become deficient in this component?*

Because all grain crops regularly remove or abstract

a certain quantity of lime from the soil, and if this is not made good the production of grain is reduced, and the straw partakes of the deficiency. This want of lime acts injuriously on man, who feeds on the grain, and on the beast, who consumes the straw.

29. *How is this deficiency made apparent?*

In man, woman, and child the bones are deficient in size, solidity, and strength; and this is still more notably apparent in all Indian cattle and horses.

30. *Is there any proof to show that the natives of Hindoostan feel this deficiency, and do what they can to make it good?*

Yes; the vast population inhabiting the plains of India habitually and constantly eat small quantities of lime paste (*chunam*) daily. The eaters of *Paun*, or betel pepper leaf, apply some lime paste to each leaf before making it up for consumption, and the chewers of tobacco leaf as a rule mix lime paste with the crumbled dry tobacco leaf before enjoying a chew. This shows conclusively that lime is needed by millions, and being deficient in most food grains,[1] and all but absent in rice, accounts for the strange habit of the people of India within and beyond the Ganges.

31. *What is magnesia?*

Magnesia is the white, nearly tasteless, powder sold in the shops under the name of calcined magnesia. It is a component of all yellow limestones of the Himalayas. In the Madras Presidency magnesian limestones are very common. In the plains of the Bengal Presidency we have no magnesian limestone; but a very efficient

[1] The very fowls lay eggs with soft shells whenever the wheat given to them is deficient in lime and lime phosphate.—*J. F. P.*

substitute is met with in the mineral called steatite, or soap stone. It is the *Sael khurree* of the Punjab Bazars, and the *Sung jurras* of Agra, Delhi, and the North-Western Provinces. It is largely produced in the Gwalior territories. One hundred parts of steatite contain, of magnesia 44, of silica 44, of alumina 2, of iron 7·3, of manganese 1·5, of chrome 2, with a trace of lime. Magnesia is present in all bones (fresh as well as fossil), and 100 pounds weight of the ashes of wheat and Indian corn contain respectively 12 and 16 pounds of magnesia, and it is also present in all kinds of grain. This will show that magnesia is needed by all crops, and that it is a component part of all superior, rich, and fertile soils.

32. *What are oxides of iron?*

"When polished iron is exposed to the air it gradually becomes covered with rust. This rust consists of the metal iron, and of the gas oxygen which the iron has attracted from the moist air, and hence it is called an oxide of iron."—*Johnstone.* The vernacular name for this oxide is *Loohae ka-moorcha.* There are two oxides of iron, the red and the black; the first forms common rust, and the scales which fall from the anvil of the blacksmith represent the second.

33. *How can it be proved and shown that iron is present in the soil?*

When a brick, or any clay vessel made by the potter, is fired and baked, it changes to a red and red-liver colour: this is due to the presence of iron. But if a vessel made of white Delhi clay be fired, it does not become red, because this clay does not contain red oxide of iron.

All red, reddish, liver-coloured, yellowish, and black soils contain iron, the last being caused by the presence of natural black oxide of iron in the soil. If common salt in powder be mixed with sulphate of iron (called the protoxide of iron), also in powder, the colour of the compound becomes yellow, thus proving that the iron present in yellow soils has been acted upon by sulphur and chlorine at some very remote period of the earth's history.

34. *Is oxide of iron present in all kinds of grain consumed by man, and in all kinds of straw, hay, grass, and fodder eaten by animals?*

Yes. All food grains, grasses, and fodder contain oxide of iron, and it is thus introduced into the blood of human beings, birds, and beasts, and it is essentially necessary to keep one and all strong and healthy. The blood circulating in the system of a well-developed adult European contains 100 grains of iron in solution. The blood of the races inhabiting India contains less.

35. *What is oxide of manganese?*

Oxide of manganese is a substance consisting of oxygen and a metal called manganese. It is very like oxide of iron, but occurs in soils and plants in very small quantity. The *Kala-muckkole, i.e.*, black *muckkole*, of the Himalayas, is well known to all hill men, and it contains oxide of manganese and iron in combination. Good tea contains manganese as well as iron.

36. *What is potash?*

When wood, charcoal, hay, and dry weeds, &c., are set on fire, they burn away until nothing but their ashes are left. When these ashes are collected and placed in a vessel, and water is poured over them, all their soluble matter is taken up by the water, which has now a peculiar taste, called in English alkaline (from the Arabic *Al-kali*), and in Hindoostanee *Kharee*.

When the solution of wood and other ashes is boiled away till all the water disappears the substance left at the bottom of the vessel is called potash in English, and *Rauk-kee-khar* in Hindoostanee.

37. *In what saline substances is potash always and largely present?*

In saltpetre or nitre, which is a compound of potash and nitric acid, 101·3 parts of nitre contain 47·15 parts of potash and 54·15 parts of nitric acid. Its Hindoostanee name is *Shorah*, and its contained acid is called *Shorae-ka-taiz-aub*.

38. *What is soda?*

In India soda is made from *Rhae*, and also from the ashes of the *Lana*, or soda plant, which is so called in Punjabee. The manufactured soda is in Hindoostanee called *Sujjee*, and is used by the Dhobees for washing clothes.

39. *In what saline substances is soda always and largely present?*

First. In common culinary salt, ten seers of which contain four seers of pure soda united to six seers of a peculiar gas called chlorine.

Second. In nitrate of soda, 85·45 parts of which contain 31·3 of soda and 54·15 parts of nitric acid. It is a natural product of India.

Third. In sulphate of soda, 71·4 parts of which contain 31·3 of soda and 40·1 of sulphuric acid.

This salt is largely manufactured in India, and is called *Kharee neemuck*.

40. *What is chlorine?*

"Chlorine is a kind of gas, which has a greenish-yellow colour, and a strong suffocating smell, and is two and a half times heavier than common air. A taper burns in it with a dull, smoky flame, and animals die in it. It exists in common salt in large quantity."—*Johnstone.*

41. *What is sulphuric acid?*

"Sulphuric acid is a very sour, burning, heavy, oily liquid, which becomes hot when mixed with water. It is manufactured from burning sulphur."—*Johnstone.*

42. *In what substances is sulphuric acid naturally present?*

It exists in sulphate of lime (Urdu name, *Sheereen-guch*), in alum (Urdu name, *Phit. kurree*), in sulphate of soda (Urdu, *Kharee numuck*), in sulphate of copper (Urdu, *Neela too-tee-ah*), and in sulphate of iron (Urdu name, *Heera kussees*). All these substances are of value to the agriculturist. The Urdu name for sulphuric acid is *Ghundhuck-ka-taiz-aub.*

43. *What is phosphoric acid?*

Phosphoric acid is a very sour, solid substance, which is formed by burning phosphorus in the air. It exists in large quantity in the bones of animals, as well as in fish bones. It is present in oyster shells, and in all fossil bones, as also in certain limestones. Such fossil bones and limestone are most abundantly met with in the Sewallic range of hills, within one day's march of Dehra Doon.

44. *What remarkable phenomenon or occurrence takes place when phosphorus is burned?*

If one hundred seers of dry wood be carefully burned to ashes, not more than one seer of wood ashes will be left; but if one hundred seers of phosphorus be properly burned, the result of the conflagration is two hundred and twenty-seven and a half seers of phosphoric acid.

Burning phosphorus gives off a white smoke, and this white smoke is phosphoric acid. The increase of weight is due to the absorption of oxygen gas from the atmosphere.

45. *What is phosphorus?*

Phosphorus is a non-metallic element obtained from bones; in appearance it is semi-transparent, colourless, wax-like, solid. It emits white vapours when exposed to the air; when ignited it burns with a luminous flame, and produces dense white fumes, which are easily collected, and, as stated above, constitute phosphoric acid. Phosphorus has a very peculiar smell, and is made in immense quantities to meet the wants of the lucifer match trade.

"The bones of a full-grown man contain 1 to 1 ⅓ pounds of this same phosphorus."—*Johnstone.*

46. *In what vegetable products is phosphoric acid always present?*

It is present in all food grains, and in wheat, barley, and other kinds of straw, also in all cultivated grasses, as well as in *Doob* grass (*Cynodon dactylon*), and other superior Indian grasses.

One hundred seers of the ashes of the grain of wheat contain 46 seers of phosphoric acid, whilst only 5 seers of it are present in 100 seers of wheat straw ashes. The same quantity of Indian corn (*Muckkee*) ash contains 45 seers of this acid.

47. *Is this acid present in the pure state, or in combination with something else?*

In all kinds of food grain it is always present in combination with saline and mineral matters, to which the name of food phosphates has been given. They are called the phosphates of soda, potash, lime, magnesia, and iron.

48. *What is carbonic acid gas?*

Carbonic acid gas is a kind of air which is without colour, but has a peculiar smell and a slightly sour taste. Burning bodies are extinguished by it and animals die in it.

49. *How can its existence be shown and proved?*

If some lime (*Chunam*) be placed in a glass, and some vinegar or diluted sulphuric acid be poured over it, immediate commotion will take place, and numbers of bubbles will form and burst and give off their contained air, which is carbonic acid gas, and is derived from the lime (*Chunam*) in which it was present in the fixed state until released or set free by the action of the vinegar or sulphuric acid.

50. *What have the soil, the air, and growing plants and trees to do with this acid gas?*

Its presence in the soil is generally due to the decomposition of all kinds of organic matters. It is always present in the air in small quantity, for in 5,000 gallons of air there are only two gallons of carbonic acid gas. All growing plants and trees feed on this gas—that is, they drink in a very large quantity of it, all of which is derived from the air.

51. *What does carbonic acid gas consist of?*

Carbonic acid gas consists of carbon (or charcoal) and oxygen. For six seers of carbon and sixteen seers of oxygen form twenty-two seers of carbonic acid gas.

52. *How can this be proved?*

By burning charcoal in oxygen gas, when carbonic acid gas will be formed.

53. *How can plants drink in so large a quantity of this gas from the air if the air contains so little?*

"They spread out their broad, thin leaves in great numbers through the air, which is always in motion, and are thus able to suck in the carbonic acid gas, from a large quantity of air at the same time."—*Johnstone.*

54. *How do they suck it in?*

"By means of a great number of very small openings or mouths, which are spread especially over the under surface of the leaf."—*Johnstone.*

55. *How do the leaves of water plants, such as the lotus* (Kownl gutta), *water nut* (Singhara), *and numerous others which grow on the water's surface, suck in this gas?*

The leaves of such plants have their mouths placed over their *upper surface*, and are thus enabled to drink in the gas needed for their development.

56. *Do the leaves suck in this carbonic acid gas at all times?*

No; only during the daylight. During the night they give off a quantity of carbonic acid gas. It is owing to this occurrence that the air of Indian forests and jungles is so unwholesome and dangerous to inhale at night.

57. *In what form do plants take in carbon from the soil?*

"In the form of carbonic acid, humic acid, and some other substances which exist in the black vegetable matter of the soil."—*Johnstone.*

58. *Do plants feed on anything else derived from the soil and air?*

Yes, on three gases, called hydrogen, oxygen, and nitrogen, and watery vapour.

59. *What is hydrogen?*

Hydrogen is a kind of air or gas which burns in the air, and it is 14½ times lighter than air. When mixed with common air it will explode with a loud report if brought near the flame of a candle. When united to oxygen it forms water.

60. *What is oxygen?*

Oxygen is also a kind of air or gas. A candle burns in it with great brilliancy; animals may inhale it for a short time without danger to life. It is 16 times heavier than common air. It is the principal component of water.

61. *What is nitrogen?*

Nitrogen is a kind of air different from the other two. It does not take fire when brought near the flame of a candle. It is injurious to life, and may not be inhaled by itself. It is a little lighter than common or atmospheric air.

62. *Does the air we breathe contain oxygen and nitrogen gases?*

Yes. Ten gallons of common air generally contain two gallons of oxygen and nearly eight gallons of nitrogen.

63. *In what form do plants take nitrogen from the soil?*

In the forms of ammonia and nitric acid. The smell of the cow-house and stable is due to the former, and the latter is present in saltpetre and nitrate of lime. If a green tobacco leaf be available and it be rubbed down to a paste in a mortar, and some lime (*Chunam*) be added thereto, and the two mixed together, gaseous ammonia will at once be given off, and may be detected by its pungent smell. If a feather dipped in vinegar or hydrochloric acid be held over the mixture, a white vapour will at once appear, thus demonstrating to the eye that ammonia is being given off, and also proving that it was derived from the tobacco leaf, which had absorbed nitrogen from the atmosphere.

64. *What is nitrate of lime?*

The fine white salt which forms on the walls of some *pucca* buildings is nitrate of lime, called in Urdu *Choonac-ka-khar.* It is often present in burned bricks and in many mud walls, from which it is scraped and collected by the saltpetre makers. Parts 82·65 of nitrate of lime contain 54·15 parts of nitric acid and 28½ parts of lime. Like potash and soda saltpetre, it is of decided value as a manure.

65. *What is watery vapour?*

If some water be placed in a pan, and the pan be placed in the sun, the water will gradually disappear, being converted into watery vapour. Dew is watery vapour, deposited by the air on the leaves of plants and on grass.

Table showing Composition of Soils of Different Degrees of Fertility.

COMPONENTS.	FERTILE WITHOUT MANURE.	FERTILE WITH MANURE.	DECCAN COTTON SOIL.	FRANCE, CANNES VALLEY SOIL.	BARREN SOIL.
Silica and Sand	648·00	833·00	482·00	860·00	778·00
Alumina	57·00	51·00	203·00	30·70	91·00
Silicate of Alumina ...	...	...	...	13·00	
Silicate of Magnesia ...	...	...	...	8·50	
Organic Matters	97·00	50·00	43·00	22·30	40·00
Carbonate of Magnesia ...	8·50	8·00	102·00	7·00	1·00
Carbonate of Lime ...	59·00	18·00	160·00	6·80	4·00
Sulphate of Lime	...	...	...	5·80	
Phosphate of Lime and Magnesia	...	...	...	5·60	
Potash Salt	2·00	Trace	...	4·80	Trace
Soda Salts	6·00	Trace	...	6·70	Trace
Carbonate of Iron	61·00	30·00	10·00	10·50	81·00
Carbonate of Copper ...	...	...	...	1·70	
Oxide of Manganese ...	1·00	3·00	...	...	0·50
Sulphuric Acid	2·00	0·75	...	...	Trace
Phosphoric Acid	4·50	1·75	...	...	Trace
Carbonic Acid with Lime and Magnesia ...	40·00	4·50	...	...	Trace
Loss during analysis ...	14·00	...	...	16·00	4·50
Total	1000·00	1000·00	1000·00	1000·00	1000·00

Explanation of the table showing the composition of soils of different degrees of fertility.

The first, second, and fifth columns show the composition of certain soils in England, of which Professor Johnstone states as follows, viz.: "The soil of which the composition is given in the first column had produced crops for sixty years without manure, *and still contained a sensible quantity of all the substances required by plants.* That of the second column produced good crops when regularly manured. *It was in want of three or four substances only, which were given to it by the manure.* The third (fifth column of the table) was hopelessly barren. *It was in want of many substances which ordinary manuring could not supply in sufficient quantity.*"

The third column of the table gives the composition of a very fertile soil in France, the lands, which are extensive, being devoted to the cultivation of numerous varieties of sweet-scented flowers, including the sweet-scented Indian *Babool* (*Acacia Farnesiana*), whose flowers when distilled yield a delicious perfume. All these flowers are needed for the manufacture of perfumes, by Monsieur Lubin, the well-known perfumer of Paris and London, who is the proprietor of the estate.

The fourth column gives a rough analysis of the famous Deccan cotton soil. Its richness in the carbonates of lime and magnesia indirectly show that phosphates of lime and magnesia are present, and potash and soda cannot be absent, otherwise the cotton plant, which needs potash and phosphates, would not grow so luxuriantly on this very remarkable and peculiar soil.

The very large quantity of lime present in this soil shows conclusively that no injury is likely to result from the use and application of lime (even in large quantity) in its uncalcined state to any soil, whilst those deficient in lime would be greatly benefited. Magnesian limestone in the same state acts beneficially, as I have seen in the vicinity of Simla luxuriant wheat crops growing on land covered with fragments (all sizes) of magnesian limestone undergoing disintegration from frost. Calcined magnesian limestone would, in large quantity, no doubt act injuriously.

Mr. C. S. Faddy, Engineer, Hoshiarpore, Punjab, applied one maund of bone-dust to three-quarters of an acre of land which was sown with cotton seed in drills four feet apart, and obtained "a few seers over ten maunds, or in round numbers, 800 lbs. of cotton in seed." The yield indicates that the soil was rich in lime. Mr. Fuller, Superintendent, Model Farm, Cawnpore, obtained "*nearly one hundred and sixty-two pounds* of cotton to the acre." Seed cotton? The soil must be very poor in lime.

CHAPTER II.

PLOUGHING AND PREPARING THE GROUND FOR SOWING SEED.

66. *How is arable land prepared for sowing?*

In the plains, generally by ploughing; and sometimes by digging with the mattock or *Phourah*, and in the Himalayas, in addition to ploughing, by breaking up the soil with an iron implement about four feet in length, half an inch (sometimes more) in thickness, and an inch and a half in breadth, and having a chisel end an inch and a half wide, which most useful implement is called *Maind* by the Paharees, or hill zemindars, who rarely use the *Phourah* in localities which, owing to steepness or inaccessibility, cannot be ploughed up.

67. *What is the time occupied by the cultivation of crops in the hills (Himalayas) and in the plains?*

In the hills, two crops are raised within twelve months. The wheat and barley sown in October — November rapidly germinates, and the young growing crop remains under the fallen snow from December to March following. These crops are harvested in May and June, according to altitude above the sea-level.

The system of harvesting is peculiar. The ears of corn with about ten inches of stalk are cut off with a small sickle; the rest of the crop is left standing to get dry, and when dry is systematically set on fire, which is the first step taken towards preparing the land for sowing the maize, millet, and other rain crops.

The second part, or ploughing, takes place after the first fall of rain, the rice crops being sown at very much lower elevations when the rains set in.

In the plains, two crops are also sown, raised, and harvested within twelve months, with the exception of sugar-cane, which is cut for the first time in the second year of its growth in October, and thereafter at discretion.

The wheat, barley, and other cold weather crops are sown in the plains from October to December, according to latitude. At harvest the corn crops are cut down as close to the ground as possible, and the straw is as carefully removed as the grain; hence nothing is left for burning. The harvest takes place from March to May, in Upper India and the Punjab. In Behar and Benares it takes place earlier.

In the plains, after removal of the harvest, the lands not under canal irrigation are left untouched till the rains set in, when those intended for rain crops are ploughed up and sown, whilst very little land is left in fallow. In the hills, wheat and barley lands are often left in fallow, as maize and millet crops will not grow at elevations suited to wheat and barley.

68. *If the land is covered with weeds or long grass, what should be done?*

In the Himalayas, all fallow land gets overrun with weeds and long grass. The zemindars enter the fields eight to twelve days before ploughing, and with long sticks break and beat down the weeds, long grass, &c.,

and allow them to dry. The land is then ploughed for the first time for wheat and barley crops. The weeds, &c., are then collected into small heaps, and set on fire, and allowed to burn out. The ashes, charcoal, and charred substances are then spread over the surface of the soil, and act as a manure. The second ploughing takes place after the first fall of rain in October; the land is then manured, rolled, levelled, and prepared for sowing wheat and barley, which is done as soon as the ground is ready.

The maize, millet, and other crops having been removed, the lands on which they were grown are also prepared for sowing, the maize and millet stalks left in the field being burned. Poppy is the only crop sown at the same time as wheat and barley. The plants are fairly grown before the snow falls, and the opium is all collected in May.

In the plains, the hot and dry season sets in after the zemindar has harvested his wheat, barley, gram, dall, and other cold weather crops; the ground becomes hard and sun-baked, weeds and grasses almost dry up, and, save where the land is under irrigation, nothing green is to be seen. The ground cannot be ploughed up owing to its hardness, and the zemindar has to wait till the rains commence, when he ploughs the land to the best of his knowledge and ability for his maize, *Jowar* (*Holcus sorghum*), and various millet crops, and other inferior grain crops, which, good, bad, or indifferent, are harvested in due course, not a thing being left standing in the field which can be removed. Hence the hill zemindar's practical and valuable system of burning stalks, straw, weeds, &c., and so restoring twice a year to the soil a certain sensible portion of the fertilizing substances named in the first part of this work, and taken up by various crops, and so removed from the soil, is not adopted by the zemindar of the plains, who thus annually impoverishes his lands, and is greatly surprised at the soil's deterioration, and at the yield of his grain crops becoming annually smaller in quantity and very often defective in quality.

69. *What should be done by the zemindars of the plains to check, arrest, and remedy this serious evil and defect in their agricultural operations?*

They should be made to understand the parable of the man and the purse of money. The owner kept on annually taking something out of it, but never thought of putting anything into it. The purse was a long one and well filled when he got it. But, to the owner's intense surprise, after many years he came to the bottom of his purse, and found it empty, and from being well-to-do became poor, and was finally reduced to the position of a Free-man in name and a slave in reality.

EXPLANATION.

The well-filled purse is the tract of fertile land inherited from zemindar ancestors. Its cultivation year by year without any rest, and suitable manuring, and removing the grain, and all straw, represents continuously taking money out of the purse, and putting nothing into it. The exhaustion of the purse represents the slow and sure exhaustion of the soil, and the zemindar's position is that of a Free-man in name who is in reality the slave of the man who feeds him, to keep himself and his family from starvation, and so holds him in bondage and hopeless debt.

The parable being explained, the next thing to be done is to remedy the evil and to remove the defects. The plan is simple and inexpensive. The zemindar should leave one-third of the straw of the wheat and barley crops standing, after removal of the ears of corn, with the other two-thirds, and when dry set fire to it, and so manure the soil with the ashes, which contain the mineral components of the straw.

The heat of the fire will roast considerable portions of the surface soil of the field, and so produce a well-known fertilizer in the shape of burned or roasted earth. The maize and millet crops sown on the land so treated will be far more productive than before. In like manner the zemindar should leave two feet in height of the stalks of these crops standing, and when dry set them on fire and plough in the ashes. The wheat, barley, and other cold weather crops sown on lands so treated, it having been twice enriched with ash manure and roasted earth, will yield far more grain than the same extent of land not so treated. Hence the temporary loss of some maunds of straw, and maize, and millet stalks will be amply compensated by the very considerably increased yield of grain.

The zemindar should keep on repeating this process year by year, and the gradual but steady enrichment and improvement of the soil will take place, and by applying the manures named in the next chapter of this book to the growing crops as directed, further improvement will result, and the yield of grain greatly increased, if not doubled.

70. *What should be done by the zemindar in the plains*

during the rains, when weeds and long grasses are so abundant, and field weeding operations have to be carried on?

All weeds and grasses which cattle will not eat should, if the shelter of a thatch be available, be dried for burning. If cover be not available, all weeds, &c., with their roots should be chopped small, with a *Phursa*, or zemindar's straw chopper, and stored in a covered pit to dry and decompose, which they will very speedily do, and so become ready for use as component of a compost, or compound manure. The leaves and stalks of such plants as the *Muddar* (*Asclepias gigantea*, Linn.), "Camel's thorn" (Hindee, *Jowassa;* botanical name, *Hedysarum alhagi*), *Bhungra* (H.) and *Bhut-kutaieah* (H.) and *Dhutoora* (H.), *Datura stramonium* or thorn apple, should also be collected and similarly disposed of.

71. *How should the land be ploughed?*

On this important subject no hard-and-fast rule can or should be laid down. The zemindar and his ploughmen are, from carefully acquired experience and knowledge of the soil, and subsoil, far better able to decide this question than men who have never seen it.

The object the zemindar has ever in view is to impress on his ploughman that on no account is he to break up what in India corresponds to the *pan* (and called in Hindee *Towa*), in the act of ploughing. Hence it follows that if the *Towa* be three, five, seven, or nine inches below the surface of the soil, the ploughing must be regulated accordingly. Therefore what is derisively called "scratching the surface of the soil," by Europeans unacquainted with its unknown and unseen peculiarities, is by no means due to ignorance, but to a knowledge of two valuable facts, the first being a sensible and detrimental decrease of produce if the *Towa* be disturbed or broken up ; the second, that wheat and barley crops throw out their roots and rootlets in such a manner as to keep close under the surface of the soil. The root growth is at first downwards, and seldom exceeds six inches, and then the rootlets grow laterally in all directions, keeping close under the surface, and thus naturally avoiding the subsoil, which may or may not be of injurious quality.

72. *If the soil be fertile and deep, what should the zemindar do to secure the best results ?*

If he possesses a modern improved Indian plough, he should use it so as to have the soil well and speedily ploughed to a moderate but sufficient depth, if intended for wheat crops raised from indigenous seed. But if the seed wheat has been originally obtained from Palestine, Egypt, Russia, America, Europe, and Great Britain, the ploughing for such seed wheats should be deeper.

73. *Why should this difference in ploughing be necessary?*

The British farmer's system of ploughing with the common English plough drawn by two plough horses, causes portions of the soluble plant food present in the upper portion of the soil to sink downwards, and as a natural consequence the roots of the growing grain crops in Europe, and especially in Great Britain, descend into the subsoil in their search for plant food; and we are told as a fact that in the United Kingdom "the roots of grain crops, of clover, and of flax will go down three feet in search of plant food, and even turnip roots in an open soil will go down upwards of two feet."—*Johnstone.*

The root growth of Indian wheat and barley being principally horizontal in place of vertical descent, the use of the European plough is not suited to the soil and plant, for the first heavy fall of rain would send a very considerable amount of plant food beyond the reach of the roots, and heavier and subsequent downpours would wash the soil almost clean, and the yield of grain would be deficient in quantity and defective in quality.

The deeper ploughing of deep, fertile soils, recommended for acclimatized seed wheat and barley, the produce of imported seed, is necessary for their wants and mode of root growth, and this system of deeper ploughing for these cereals will have the effect of preparing the soil for sowing with imported broom corn,[1] producing 219 maunds of grain per acre as an ordinary crop, and also for the "Branching *Doura*," a remarkable variety of "Sorghum," or *Jawar*, which yields from six to sixteen stalks from each seed sown, and as many heads of grain, whereas the common *Jawar* of India produces only one stalk and one head of corn or grain.

74. *In how many ways does the British farmer plough his lands in view of their improvement, and the realization of heavy or first-class crops?*

To begin with, the British farmer, in addition to ample working capital, has not only all agricultural implements, but has different ploughs for each description of ploughing. First, he has a common horse plough of the best construction, for common ploughing. Second, he has a subsoil plough, for subsoil ploughing. Third, he owns a trench plough, for trench ploughing. Fourth, a double mould board plough, for drilling land; and fifth, a small plough, constructed like the common plough, for small and peculiar work.

[1] The "Branching *Doura*" and Illinois (U. S.) "Broom Corn" are found to be identical.

75. *How do these ploughs act upon the soil?*

First, with the common plough, drawn by two horses, farm lands are ploughed at discretion in the usual manner. This plough merely turns over the surface-soil. Second, with the subsoil plough, which stirs the subsoil deep without bringing it to the surface, thereby admitting into it rain and air, to make useful what might be hurtful in it to plants. This plough stirs and loosens the subsoil, and prepares it (subsoil) for—Third, the trench plough, which brings the subsoil to the surface and puts the surface-soil to the bottom of the trench.

For certain crops, the soil, after being duly ploughed and pulverized, has to be drilled, and for this purpose the British farmer employs his fourth plough, called the double mould board plough.

The fifth, or small plough, is used for paring away the soil from the sides of the drills. This plough is also used when the land is ribbed.

76. *Is it necessary for the Indian zemindar to adopt any of these modes of ploughing?*

The zemindar might, under certain conditions of surface-soil and subsoil, adopt subsoil ploughing, and also trench ploughing. But before he could do so it would be necessary for the European agricultural implement maker to produce a subsoil plough, as also a trench plough, which could be drawn by two camels or four large bullocks, or six of smaller size.

77. *What benefit is likely to result from subsoil ploughing?*

The air and rain would descend into the subsoil; the latter would keep it moist, and the former, in combination with the moisture, would so change the subsoil (supposing it to be in need of such change) as to gradually make it fit to be brought to the surface, which would subsequently be done by the trench plough.

78. *If the surface-soil and subsoil were both fertile, would subsoil and trench ploughing be beneficial?*

Yes. In this case the trench plough would follow the subsoil plough, and the soil brought to the surface being fertile would yield heavy crops; as all valuable soluble substances which had descended from the surface-soil into the subsoil would be brought to the surface, and so afford an ample supply of plant food for the growing crop.

79. *What crops would be benefited by this method and description of ploughing?*

It would benefit all grain crops, and especially sugarcane and cotton; as the cotton plant, which, in addition to roots near the surface, possesses a tap-root, which cannot well descend into the subsoil in search of plant food if the soil be hard and caked.

80. *Is there any description of ploughing the land peculiar to India?*

Yes. Lands to be placed under a rice crop are ploughed after a heavy fall of rain, when they are more or less under water. The soil is then quite soft and plastic, and with the light Indian plough the zemindar very speedily prepares it for rice culture. The various ploughs used by the British farmer could not be of any use for this purpose. Rice lands, under artificial irrigation, are ploughed and levelled with very remarkable ingenuity and accuracy, and are flooded and drained in the most practical and effectual manner.

81. *The British farmer employs powerful horses exclu-*

sively for all his farming operations; what quadrupeds does the zemindar employ?

As a rule, the zemindar employs the bull and ox, and sometimes the buffalo, whilst in certain parts of India the camel is used to plough the land.

82. *Would it be advisable to employ the camel more extensively for agricultural purposes?*

Yes. Before long camels will be on sale at moderate prices, and one is strong enough to draw a plough which requires two bullocks. When not needed for agricultural purposes, the camel is of considerable value as a beast of burden, carrying six maunds with ease. The female camel gives a large supply of good rich milk; and her calf may be kept for a mere trifle, and when full grown sold to advantage. The camel eats all kinds of green and dry fodder, including leaves of the *Neem* (*Azadiracta Indica*), which are so very bitter that no other animal will touch them; and finally, the camel supplies an abundant supply of very superior manure, extremely rich in ammonia; so much so, that in the district of Panneeput, formerly in the North-Western Provinces, but now in the Punjab, immense quantities of camel's dung are collected and operated upon by men engaged on the manufacture of *Sal ammoniac* (muriate of ammonia) therefrom, and who thus supply the merchants engaged in this trade.

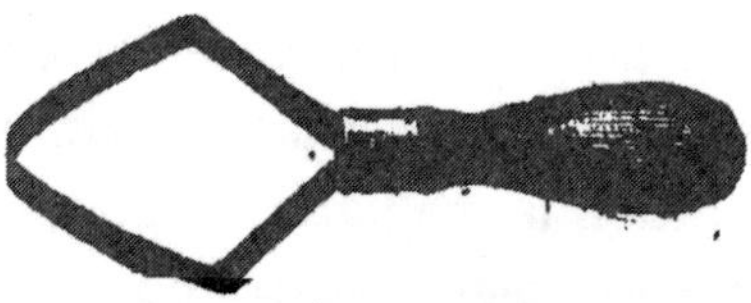

NOYES' HAND WEEDER.

CHAPTER III.

MANURES AND COMPOSTS.

Farmyard and Portable Manures.—Saline Manures.—Mineral Manures.—Metallic Manures.—Vegetable and various other Manures.

83. *What is meant by manuring the soil?*

The practical farmer collects and converts refuse matters, such as dung, urine, poultry-yard sweepings, and decomposed vegetable substances into a compound called, in English, farmyard manure, and in Hindee, *Khaud.* When this substance is applied to land, spread out thereon, and ploughed into the soil, the process is called manuring the land. The Chinese farmer adopts the more economical plan of manuring the growing plant, or manuring the circle or spot where one or more seeds are to be sown. The hill zemindar applies cattle manure by hand to his wheat and barley crops, as a top dressing after the snow is melted. The object is two-fold: neither hares nor deer will enter or eat a growing crop so protected; and the first fall of rain dissolves the manure. The solution immediately supplies the rootlets near the surface with plant food, and as it descends deeper the roots are also fed. If the weather be clear and the sun hot, ammonia is freely given off by the manure applied as indicated, and in this case the leaves of the wheat and barley plants absorb it at their leisure and pleasure, and thrive accordingly.

84. *Besides farmyard manure, are there other substances which can be and are used as manures or components of manures? and what are they called?*

Yes. They are technically called portable manures, on account of their being of small bulk or weight as compared with cattle or farmyard manure, and can easily be transported to great distances at a moderate cost.

85. *What is the advantage of portable manures over farmyard manures?*

They can be used with profit by the farmer even if brought from foreign countries or distant inland places by water carriage or railway.

86. *Does India possess any of these portable manures? If so, name them in detail.*

There is no country under the sun which possesses valuable portable manures in greater abundance and variety than does British India. But, most unfortunately for the Indian agriculturalist, the Government of the late Honourable East India Company placed all saline substances under the Excise, and the heavy duty imposed having been continued and maintained by the present Government under the Crown, such substances, on account of the prohibitive duty, cannot be used for manuring purposes by well-to-do zemindars, and are quite beyond the reach of the ryots and ordinary cultivators of the soil.

The portable saline manures are (1) chloride of

sodium, or common salt; (2) sulphate of soda; (3) nitrate of soda; (4) carbonate of soda; and (5) nitrate of potash. The portable mineral manures are fossil bones, fossil phosphate of lime, phosphatized limestone, magnesian limestone, ordinary limestone, *Kunkur*, and gypsum, or sulphate of lime. The sulphate of alumina and potash, or common alum, and the sulphate of iron are also portable manures, whilst the sulphate of copper is needed by the farmer for making *pickle* for his seed wheat.

The first five are under the Excise, and the others are neither appreciated nor used by the zemindars of India. This is due to want of information on the subject, which it is the object of this work to supply.

87. *Is common salt of established value as a portable saline manure? and if so, adduce the proof.*

Yes; the blood of man in particular, and that of all animals and birds contains a notable quantity of salt, which is derived from the food consumed by them. It is a well-established fact that "salt exists in, and is necessary to, all cultivated plants, especially to root crops, such as turnips, mangold wurzel, potatoes; to leaf crops, like cabbage; and to young shoots, like asparagus. When applied to grain crops, common salt almost always increases the weight per bushel of grain when when reaped. Mixed with fresh-slaked quicklime and put on the land, it gives strength to the straw.—*Johnstone.*

88. *In what cereals under cultivation is common salt fully present, if the land has been treated with salt as one of the components of the manure used?*

Reliable analysis shows that 100 lbs. of the ashes of wheat contain 10 lbs. of common salt; 100 lbs. of the ashes of wheat straw contain 11 lbs. and 11 oz. of common salt; 100 lbs. of barley ash contain less salt than does wheat. Potato ashes contain per 100 lbs. as much salt as wheat, whilst 100 lbs. of maize, or Indian corn ashes, contain only one twenty-fourth part of the amount of salt present in wheat ashes, or 6⅔ oz. This shows that wheat requires salt in large quantity, Indian corn or maize in very small quantity, and sugar-cane, sugar sorghum, sugar beet, and cotton in still smaller quantity.

89. *Can it be proved that land manured with common salt and nitrate of soda, and subsequently sown with wheat, gave in consequence a large and satisfactory return?*

Yes. Certain lands, afterwards sown with wheat, were manured with these salines in the proportion of 3 cwt. (4 maunds and 3 seers) of common salt, and 1½ cwt. of nitrate of soda per acre. The produce or wheat harvested was 2,804 lbs. per acre, or 47½ bushels. The adjacent acre of unmanured land gave only 2,004 lbs. of wheat, the difference or increase being 800 lbs., or 13½ bushels. The acre of land manured only with 1½ cwt. of nitrate of soda gave 2,606 lbs. of wheat, whilst the acre manured with 3 cwt. of salt only, gave 2,080 lbs. of wheat. These facts and figures show conclusively that salt and nitrate of soda in combination gave the highest yield of wheat, beating all other manures, of which seven kinds were used. The yield of straw indicated an increase of 12½ cwt.

90. *Why cannot common salt be used in India as a component of a manure?*

Because the Excise duty levied on a ton of salt

amounts to Rs. 54, or at par to 108s., and when this salt is placed in the market by the salt merchant, its price per ton is as a minimum just doubled. The British farmer can purchase as much salt as he likes at 15s. per ton, and can therefore use salt as a manure at discretion.

91. *What prevents the nitrate of soda and the far more valuable nitrate of potash from being used in India as a component of a manure?*

Both these salines exist largely in certain soils, and both are more or less naturally associated with common salt. There is no sale for the nitrate of soda, and to manufacture common salt therefrom, when so associated, by separation, would cause the salt maker to suffer a dead loss, for though he must pay Rs. 2 (or at par 4s.) per maund of 82 lbs. as Excise duty on the common salt present, the nitrous product is unsaleable. The crude nitrate of soda as well as the crude carbonate and sulphate of soda all contain more or less salt which the Excise Department looks upon as a sort of sacred condiment, not to be meddled with by any one until it has paid duty, and then the agriculturist who had in his possession more than a few pounds of salt (duty paid) would stand an uncommonly good prospect of being arrested as a smuggler. The Excise Department does not believe that common salt can be used as a manure, and considers the giving of salt regularly to cows and plough cattle as quite unnecessary; the value of cattle salt being unknown to the Salt Department, though fully appreciated and recognized in Germany, which country, like India, derives a state revenue from its salt monopoly.

The nitrate of potash is always saleable, and when

separated from common salt both may be sold, after payment of salt duty and license fee. But such saltpetre is too high-priced to be used for manurial purposes. The sulphate and carbonate of soda being also generally associated with common salt, are *ipse facto* barred to the Indian agriculturist.

92. *If nitrate of potash, or saltpetre, was cheap, would the zemindars use it as a manure component?*

Yes. And it would be found of much higher value than nitrate of soda. Saltpetre contains the largest quantity of potash in the smallest bulk and best form for use as a manure.

The grain of wheat and wheat straw contain very considerable quantities of potash, derived partly from the soil and partly from the cartloads of farmyard manure applied thereto. Hence it follows that a small quantity of saltpetre would supply more potash and nitrogen than a very large quantity of farmyard manure; whilst if lime be present in the soil or manure, nitrate of lime is formed, and this, when acted on by potash, no matter whence derived, at once forms saltpetre. Thus the value of saltpetre as a manure is permanent, economical, and beneficial in every way; for once in the soil it, so to speak, keeps on manufacturing itself.

Crude saltpetre contains common salt as well as nitrate of potash, and is therefore the very best saline manure for wheat, barley, oats, turnips, and potatoes. Italian rye grass, red and white clover, lucerne, and *Reana luxurians*, or the Mexican buffalo grass, all contain and require much potash for their proper growth and development, as also common salt. Hence if crude saltpetre be not available, common salt must be added to the nitre, and both applied in solution to the farmyard manure or lime, and pulverized roasted earth manure, before being spread over the surface of the soil.

NOTE.—The writer has used nitrate of potash, or saltpetre (costing, thanks to the high Excise duty, Rs. 270, or at par 540 shillings per ton), as a component of manures for eight years, and will continue to do so as a matter of course, there being no experimenting or doubt about the manurial value of this potassic salt; whilst other potassic salts, which are used with confidence, are of doubtful value.

93. *What is phosphate of lime, and whence derived?*

If fresh bones are burned in the open so that the air acts upon them, the substance left is called bone-ash and bone-earth. One hundred lbs. of this bone-ash contain 54·05 lbs. of lime, and 45·95 lbs. of phosphoric acid.

94. *What is fossil phosphate of lime?*

In composition it is almost the same as bone-earth, the associated lime and the phosphoric acid having become of stony hardness, as is the case with most fossils.

95. *What is fossil bone?*

At some remote period of the earth's history incalculable numbers of animals, birds, fish, reptiles, &c., have been deprived of life by some convulsion of Nature. Their entire skeletons, whole bones, and bone fragments still exist embedded in a limestone matrix, and as these bones have become solid, and also of stony hardness, they are called fossil bones in contradistinction to common bones, which are more or less hollow. The fossil bone contains all the lime and phosphoric acid present in the original bone, *plus* calcareous matter, absorbed when changing or passing into the fossil state.

96. *What is phosphatized limestone?*

The blood and flesh of all animals, birds, fish, &c., naturally contains the phosphates of soda, potash, lime, magnesia, and iron. Hence it follows that when the vast destruction of animal life alluded to took place, the carcases of all, from the mastodon down to the smallest quadruped, were engulfed and embedded in calcareous paste. When the skin, flesh, and muscles of all these underwent decomposition, the lime paste absorbed much of the gases given off, and in due time the phosphates present in the flesh of the carcases was also taken up, and became amalgamated with the calcareous paste, which thus became phosphatized; and the immense scale and extent of these operations and impregnation of the lime paste with the phosphates named will be better understood when the ossiferous deposits of the Sewallic range of hills comes under notice.

97. *Is this phosphatized limestone of greater value as a manure than fossil bone, or fresh bone-ash?*

Yes, it is of greater value, and for this reason: it contains all the food phosphates and latent ammonia derived from the flesh of the embedded carcases. Fossil bone and fresh bone-ash do not contain these important substances, and as fossil phosphate of lime is derived from disintegrated fossil bone, they are absent therefrom. The fossil dung of carnivorous animals, principally of hyenas, called coprolites, is the least valuable of this class of manurial substances.

98. *To remove all doubts, and to enlighten the uninformed, can proof be adduced to show that the flesh and*

blood of animals contains the phosphates named, as being directly derived from their food?

Yes. Fresh meat, when dried and incinerated, or burned to ashes, leaves 3½ per cent. of salts, and 100¼ lbs. of such fresh meat ash contains, according to the analysis given by the chemist Keller, the components given beneath.

Composition of the Ashes of Flesh.

Phosphoric Acid...	36·60
Potash	40·20
Lime, Magnesia, and Oxide of Iron	5·69
Sulphuric Acid	2·95
Cloride of Potash	14·81
	100·25

If out of the body carbonate of potash is acted on by hydrochloric acid, till effervesence ceases, and the solution is evaporated, the residue is chloride of potash. The chlorine has been derived from the common salt present in the food. All milk contains this salt.

The cow in milk feeds entirely on vegetable substances, and her calf, chiefly on its mother's milk.

The analysis of the ashes of veal by the chemist Staffel (after deducting common salt) given beneath, shows that 100 lbs. contained of

Phosphate of Potash	68·05
Phosphate of Soda	5·66
Phosphate of Lime	3·72
Phosphate of Magnesia	6·24
Free phosphoric Acid	15·10
Peroxide of Iron...	0·30
Silicic Acid...	0·92
	100·00

If out of the body the carbonates of potash, soda, lime, and magnesia were acted on by liquid phosphoric acid, until effervescence ceased, and the four solutions were separately evaporated, the residue would constitute artificially prepared phosphates of potash, soda, lime, and magnesia. All these were naturally present in the milk of the cow, and so passed on as food to the body of the calf, the bull, or ox of the future.

The ashes of ox blood, analyzed by the chemist, Dr. Stozel, were found to contain of

Phosphoric Acid...	14·04
Potash and Soda	59·97
Lime and Magnesia	3·64
Carbonic Acid	18·85
	96·50

The common salt and iron present in the blood had been deducted before making the analysis. The loss or deficiency per 100 parts consisted of sulphuric acid, silica, &c.

These analyses prove conclusively that if a ruminant very much larger than an elephant was embedded entire in calcareous paste, the phosphates present in the carcase must of necessity be amalgamated therewith; or, to put the matter in simpler form, if 100 lbs. of the ashes of veal were to be mixed with *quantum suff.* of freshly-made lime paste, and a petrifying chemical solution added thereto, the resulting phosphatized compound would almost exactly resemble phosphatized limestone. The difference consists in the absence of absorbed ammonia, which, with carbonic acid gas, has become latent in the fossil.

99. *Can proof be adduced that the hay and grass on which milch cows are fed contains the food phosphates present in veal ashes?*

Yes. In England pasture land is regularly manured, bone-dust being used as one of the manures, and the consequence is that the hay and grass obtained from such land is rich in food phosphates. The milk partakes of this richness, and it has been shown above that the flesh of the calf contains them in very considerable quantity. Mr. Potter, chemist, analyzed 100 lbs. of the ashes of hay, which were found to contain of

Potash	20·08
Soda	10·84
Phosphoric Acid	17·35
Lime	8·24
Magnesia	4·00
Oxide of Iron	1·82
Common Salt	5·00
Sulphuric Acid	2·10
Silicic Acid	30·00
Carbonic Acid	0·67
	100·00

NOTE.—This matter has been gone into minutely, to show the value of phosphoric acid and its compounds. In India caste prejudices will and do prevent the use of bones in any form, but no objection exists to the use of fossil bone and phosphatized limestone.

The value of the ossiferous deposits extant in the Sewallics is absolutely ignored by the Indian authorities, and it will rest with private enterprise to open up this vast neglected mineral treasure if permitted to do so.

The British farmer pays £4 per ton (27 maunds) for fossil phosphate of lime. The Indian zemindar might have it for the trouble and cost of quarrying and removal if cart-roads existed. But before this could be done the wealthy titled and untitled Talookdars of Oudh, the North-Western Provinces, and Punjab will have to act in unison, and so secure for agricultural millions this most important and valuable of manures.

The European capitalist will be greatly benefited if allowed to utilize these valuable ossiferous deposits. The manufacture on the large scale of phosphorus and all its numerous marketable medicinal and chemical compounds cannot take place without cheap lime phosphates. At present India imports every box of phosphorus lucifer matches sold. But with local-made phosphorus to hand, an immense lucifer match trade, to meet the wants of all Hindoostan, would be called into existence. Thus it will be seen that the despised fossil phosphates of the Sewallics have a commercial and chemical value, in addition to their high manurial value.

100. *State what is known to the Indian authorities on the subject of ossiferous deposits.*

"In 1834 Lieutenants Baker and Durand discovered the great ossiferous deposits of the Sewallics near the valley of Markunda westward of the Jumna River, and below Nahun. The deposit was reported, and proved to be unexampled for its richness and extent. It included the earliest discovered Quadrumana, an extraordinary number of Proboscidia belonging to Mastodon Stegodon, and Elephas; extinct species of Rhinoceros, Chalicotherium, Equus and Hipparion, Hexapostodon, Hippopotamai and Mercycopotamus, Ios and Hippohyus; the colossal ruminant Sivatherium, together with species of camel, giraffe, and new types of Bovidæ;

also species of Cerous and Antelope and Capra; Carnivora belonging to the new Genus Sivalaretos, and Enthydriodon, Felis Machairodus, Hyæna, Cania, Sutra, &c. Among the Aves, species of Ostrich, Cranes, &c. among the Reptilia, Monotars, and Crocodiles of living and extinct species, the enormous Tortoise, Colossochelys Atlas, with numerous species of Emys and Trionyx; and among fossil fish, Cyprinida and Silicisda."

In 1832 Captain Proby T. Cautley explored the Kalawala Pass,[1] and by means of blasting discovered more perfect remains than had been found by Dr. Falconer, and these included Miocene *Mammalia genera*. To Captain Cautley belongs the credit of having discovered vertebrate fossil remains *in situ* in the tertiary strata of the Sewallic hills."

The information here brought to notice was originally published in the "Journal of the Asiatic Society of Bengal" for 1832, vol. i. pages 97–249, and also in vol. ii. page 182, of the same Journal.

BONE MANURE.

101. *How is bone manure made, and how many kinds of it are there?*

There are five kinds—1. Bone dust, made by reducing dry bones to powder in a mill. 2. Bone fragments, reduced to powder by fermentation and disintegration. 3. Calcined bones, reduced to powder, called bone-ash. 4. Bones broken into small pieces or fragments and so

[1] All my inquiries made during my winter's stay at Dehra Doon have failed to find the whereabouts of this Pass. It has since been spotted, and will be explored during the coming winter months.—*J. F. P.*

used. 5. Bones dissolved in sulphuric acid, and called superphosphate of lime.

102. *Can you state how the second kind of bone manure is made?*

Yes. Fill an empty one-dozen claret case with broken bones and toss them into a small pit, and add thereto half that quantity of sand or pulverized soil, and so proceed until the supply of bones is disposed of, each boxful of bone fragments having its proper quantity or half-box of sand or soil added. Next moisten well the mixture of bones and sand or soil with cold water, and turn it all over three or four times and cover the pit or hole with a mat or straw *Jhamp*. In a short time the bones will ferment and heat and gradually crumble down into a fine powder, which, with the sand used, must be stored for future use.

The pit should be dug under shelter, so that no rain or water can get to it.

103. *Explain how the fourth description of bone manure is made.*

In England bones are crushed in mills and then sifted into the various sizes of inch bones, half-inch bones, and grit, and they are thus applied to land.

NOTE.—To give the Talookdars of India some idea of the extent to which bones are used in Great Britain as manure, it is necessary to state that "in addition to the bones of home growth used as manure, 40,000 tons (10.80.000 maunds) of bones are yearly imported, and chiefly for this purpose."—*Johnstone.*

"The ossiferous deposits of the Sewallics when utilized could supply all India as well as the London market with any quantity of fossil bones as quarried, and, in addition, the far more valuable phosphatized limestone. When the Dehra-Doon railway is made, private enterprise may perhaps be allowed to quarry and remove fossil phosphates to the plains of India, and thus lay the foundation of India's future wealth and prosperity as a great wheat, barley, maize, and sorghum-sugar producing country." —*J. F. Pogson.*

104. *Explain how the fifth description of bone manure is prepared, and how used.*

"About equal weights of bone-dust and of sulphuric acid are taken. The acid is diluted with one or two waters and poured upon the bones, and the mixture is stirred occasionally for two or three days, when the solution of the bones is completed, and the superphosphate of lime is ready for use."—*Johnstone.*

N.B. The operation must be conducted in the open air, as deleterious gases are given off.

"The acid paste so obtained may be diluted with 30 to 50 times its bulk of water, and applied directly to the land with a water-cart.

"If it be desired to store or preserve this paste for future use, it may be dried up with powdered charcoal, sawdust, or soil, and when required drilled in as bones usually are."—*Johnstone.*

This acid bone paste "when dried by admixture with more bone-dust, and laid in a heap for some time, forms an excellent preparation, which is sold under the name of dissolved bones."—*See* Johnstone's "Analysis of Soils."

105. *What do fresh bones consist of?*

They consist of gelatine or glue, carbonate and phosphate of lime, with some magnesia. The composition of bones differs as to quantity of constituents ; thus 100 lbs.

of human bones contain 53·04 lbs. of phosphate of lime, 11·3 lbs, of carbonate of lime, 1·16 lbs. of phosphate of magnesia, and 1·20 lbs. of soda, with a very little salt.

The bones of the ox contain in 100 lbs. 57·35 of phosphate of lime, 3·85 lbs. of carbonate of lime, 2·05 lbs. of phosphate of magnesia, and 3·45 lbs. of soda, and common salt.

This analysis shows that the bones of the ox will yield the best manure for all grain crops, and especially wheat, wheat-barley, maize or Indian corn, and *Sorghum vulgarens—Jowar.*

The bones of sheep and poultry will make the best manure for oats, beans, peas, and *dalls* of all kinds, as also for *colza*, linseed, *till* or sesamum, and all oil-seed crops, including turnips. 100 lbs. of sheep bones contain 80 lbs. of phosphate of lime, and 19·3 lbs. of carbonate of lime; whilst 100 lbs. of poultry bones contain 88 lbs. of phosphate of lime, and 10·4 lbs. of carbonate of lime.

The bones of the frog and fish will produce the best manure for rice, barley, small millets, potatoes, carrots, and other root crops. 100 lbs. of frog bones contain 95·2 lbs. of phosphate of lime, and 2·4 lbs. of carbonate of lime; whilst 100 lbs. of fish bones contain 91·9 lbs. of the phosphate of lime, and 5·3 lbs. of carbonate of lime.

In all parts of India, where frogs and small unedible fish and freshwater shells of all kinds are abundant, they should be freely used as manure. The fish and frogs may be caught in hand nets and thrown for a few minutes into a tub or cask containing lime-wash (*Chunam* and water), which will at once kill them; after which they ought to be mixed with pulverized soil, or *Kunkur* dust, or lime riddlings, or a mixture of all three, and be buried in a pit to decompose and undergo amalgamation. Fish manure is to be made in the same way. The lime (*Chunam*) should be sprinkled over every three inches in depth or layers, of fish and frog alike; it will hasten their decay and enrich the manure.

When the decomposition has been completed, the contents of the pit should be taken out and run through a sieve, and the bones set aside to be pounded and added to the sifted manure, which should now be stored for use.

On the Malabar and Coromandel coasts vast quantities of fish may be caught and made into valuable manure in the above manner.

All freshwater shells should be similarly treated to kill the fish inside, and in due course the shells would be reduced to powder and mixed with the sifted soil. The common Indian *Dhain-Kee*, used in Bengal for pounding broken bricks into brick-dust, will answer admirably for this purpose.

NOTE.—A consideration of the above information will, it is hoped, convince the most sceptical reader in and out of Hindoostan that there is no scarcity in India of bone manures. The Hindoos, as a rule, object to the use of bones as manure, and the secretary of the Bijnor Agricultural Society, Mr. P. Sri. Lall, has done much to overcome this antipathy to the use of bone manure. But, most extraordinary and strange to say, a reference made to Mr. E. C. Buck, C.S., and Secretary to the Government of India, in the Department of Revenue and Agriculture, has elicited the extract given beneath, and which will be found in the printed Proceedings of the Bijnor Agricultural Society, of the 7th February,

1882, a copy of which is supplied to the writer as an honorary member of the Society.

The secretary states that the following suggestions have been received from Mr. Buck in reply to his memorandum.

Mr. Buck observes: "*Bone Manure.*—I believe that further experiments are much required to elicit its real value. I would suggest having experiments tried for two or three seasons (under the advice of the Cawnpore department), by members of the Society, at the conclusion of which you will be able to deal with the issue of the financial value of this manure. It will be time enough then to press its value upon others."

The following anecdote and extract from the Proceedings of the Agricultural Society of India, of the 27th of April, 1882, speaks for itself; and the former shows what little need there is for wasting time in India on bone manure experiments, as advised by the head of the Indian Agricultural Department, whilst the extract settles pretty conclusively the financial value of bone manure.

Professor Johnstone, the leading agricultural authority of England to the day of his lamented death, states, in reference to the opposition offered, by those who knew no better, to the use of bones as an agricultural manure, that "It is only about sixty years since bones began to be much used in England. When, fifty years ago, it was proposed to use them in Wigtownshire, the objection was, 'They'll do naething but breed maggots to eat every thing off the land.'"

The writer has used bone manure regularly from 1874 to 1881, and can bear witness to its very great value, as also to the fact that it did not breed maggots, which were so abundant in hill cattle manure as to require their destruction before it could be used at all.

Mr. William Claxton Peppe, of Birdpore, Goruckpore, reports: "The wheat barley promises to be a great success. It was the first ripe and first cut of any *Rubbee* crop about here. The standing crop was a magnificent sight. The produce from four-fifths of an acre manured with bone-dust was 1.156 lbs. (19 bushels and 16 lbs. of 60 lbs. each[1]) of grain, and of straw 1.640 lbs. The produce from two-fifths of an acre manured with ordinary farmyard manure was 255 lbs. (two hundred and fifty-five pounds) of grain, and 410 lbs. of straw."

To convince the critical reader of the value of Mr. Peppe's facts, it is only necessary to state that the wheat-barley manured with bone-dust produced 1.445 lbs., or $24\frac{1}{12}$ bushels of grain, whilst the acre of land manured with ordinary farmyard manure produced only $637\frac{1}{2}$ lbs., or $10\frac{15}{24}$ bushels of wheat-barley. The difference, $807\frac{1}{2}$ lbs., represents the financial value of bone-dust manure.

After this testimony to the value of bone manure nothing further need be said on the subject. The wheat-barley alluded to has also been grown by the writer, at Kotgush, in 1880–81, and measured by bulk was sensibly heavier than the same measured bulk of first-class wheat. Further information on this subject will be found under the part of this work headed "Wheat-Barley."

A bushel of the best English barley weighs 46 lbs. The weight of the wheat barley has been calculated at 60 lbs. to the bushel by the producer.

106. *Why may not superphosphate of lime be made and used in Hindoostan?*

[1] The bushel of common or husk barley weighs 46 lbs.

This compound cannot be made without sulphuric acid, which in India costs over one shilling, or eight annas, per pound, at which rate the manure would cost far more than the value of the grain crop harvested. For the same reason the fossil phosphates cannot be used in this particular manner.

107. *Is it not quite feasible, by means of simple catalytic chemical action, to convert fossil phosphates of lime into mineral manure without the use of the costly sulphuric acid?*

Yes. A very simple, effectual, and inexpensive process for so doing was discovered in 1878–79, and the fact (the *modus operandi* was not given) was reported in January 1882, for the information and consideration of His Excellency the Viceroy and Governor-General of India in Council. The report was submitted by Mr. E. C. Buck, Secretary to Government Agricultural Department, to Mr. Duthie, Superintendent, Government Botanical Gardens, Saharunpore, and with his reply the matter ended. The question was one of very considerable imperial, financial, commercial, and agricultural importance, and its suppression by the head of the Agricultural Department is to be regretted.

The formula for converting all kinds of fossil phosphates into mineral manure without the use of sulphuric acid is withheld till time and favourable circumstances will admit of its being published for the use and benefit of India's agricultural millions.

NOTE.—The official correspondence on this subject will be found in the appendix.—*J. F. Pogson.*

108. *In England why do dairy pastures especially require bone manure?*

"Because milk and cheese contain bone-earth, and if these be carried away and sold off the farm for a number of years, the land is robbed by degrees of this bone-earth more than of any other substance. Only those grasses can then grow which require comparatively little bone-earth."—*Johnstone.*

In England, every ten gallons of milk contain nearly half a pound of bone-earth. A cow, therefore, which gives twenty quarts a day takes nearly 2 lbs. of bone-earth from the soil every week. To return these 2 lbs. to the soil, 3 lbs. of dry bones or 4 lbs. of common bone-dust are required.

Again: every 100 lbs. of cheese contain about 2½ lbs. of bone-earth, requiring 5 lbs. of bone-dust to replace.

109. *In England, what effect follows from adding bone manure to old dairy pastures?*

"The bones supply bone-earth of which the land had been robbed. New or more healthy grasses then spring up which contain much bone-earth, and these when eaten by the cow produce milk in greater abundance and richer in cheese."—*Johnstone.*

NOTE.—The critical reader will not fail to observe that the action of bone manures on soil and grasses are rapid and simultaneous, and act beneficially on the cow and her milk. All the grain and root crops cultivated by man require bone-earth (no matter whence derived) for their perfect development. This subject is so well established and so thoroughly understood that neither empirical experiments or advice need be resorted to or followed if given.

The necessity for rapidly and effectually renovating the more or less exhausted arable lands of India with the fossil phosphates alluded to is undeniable, and it is to be hoped that the Indian press when criticizing this work will not fail to note the stumbling-block which exists in the pathway of India's agricultural improvement.

All who have it in their power should explain to ryots and zemindars that, as bone-earth exists in cows' milk which is the favourite beverage of the Brahmin, who eats fine flour *Chuppatties*, in which bone-earth is also naturally present, to object to its use as a manure whilst all classes consume it as part of their food is an absurdity.

CALCINED LIMESTONE, OR LIME MANURE.

110. *How is lime manure made?*

In England, limestone is burned or calcined until quicklime is produced, and this when slaked is reduced to a fine powder, in which state it is applied to the surface of the land, and worked into the soil in any convenient way.

For India this system is quite unnecessary. By the English plan the carbonic acid gas, so much needed by plants, is driven off by igneous action, and it takes months, if not years, before this gas is re-absorbed from the soil and atmosphere.

By the author's plan the carbonic acid gas is retained in the limestone, and *Kunkur* used as manure, and the growing plant derives the full benefit of both lime and carbonic acid gas.

POGSON'S PLAN FOR MAKING LIME MANURE.

In the plains, *Kunkur* when calcined yields lime or *Chunam;* hence all *Kunkur* which will burn into lime is fit for use as manure. The *Kunkurs* which will not yield lime when burned are only of use as road metal, and are unfit for agricultural purposes.

The selected *Kunkur* is to be roasted, weeds, dry grass, and some sticks, and cow-dung fuel (*Oop-la*) being used for this purpose. A layer of such mixed fuel should be spread on the ground within the circumference of a circle marked out on the ground. On the fuel place two or three inches of *Kunkur* nodules, and thereon two inches of fuel, and so on alternately till the pile is eighteen inches in height, when it should be fired all round, the object being to *roast* and *not to calcine* the *Kunkur*. When the pile is cold, rake out the *Kunkur* from the ashes, which store for use. Next reduce the roasted *Kunkur* to powder in the *Dhain-Kee*, and store for use. This substance plays a most important part in the preparation of composts, and if the soil be poor in lime or sandy, the defect may at once be remedied by applying roasted *Kunkur* dust to the soil, in the proportion of one chittack to each square yard of land, or seven maunds—twenty-two and a half seers—to the acre. If this *Kunkur* dust be repeatedly wetted with the urine of the cow, which is always very rich in potash and lime, chemical reaction will take place, and the urea of the urine will give up its nitrogen to the lime dust, and nitrate of lime and saltpetre will be formed. In this simple and inexpensive manner every zemindar may supply himself with mineral manure rich in lime, potash, and saltpetre, or the nitrate of potash.

If the roasted *Kunkur* dust be wetted with a solution of sulphate of soda, or *Kharree neemuck*, the sulphuric acid thereof combines with the lime, forming a sulphate of lime, and the soda changes into its carbonate. The mineral manure so obtained will be of considerable value for such crops as need soda and sulphate of lime.

If the mineral manure so prepared be dried and wetted with cow's urine a new combination takes place, and the manure will be rich in potash, soda, lime, nitrate of lime, and sulphate of lime.

One hundred seers of roasted *Kunkur* dust will require 5 seers of sulphate of soda (*Kharree neemuck*), which should be dissolved in 21 seers of cold water, and the solution applied to the *Kunkur* dust through the rose of a watering-pot, or, if unobtainable, gradually sprinkled thereon by hand.

111. *How are composts made?*

COMPOST No. 1.—To make a compost, to every ten maunds of farmyard manure add one maund of lime manure, obtained by the use of sulphate of soda.

COMPOST No. 2.—If it be desired to make a compost of *Kunkur* dust, acted on by cow's urine as explained, then to each maund thereof add 15 maunds of farmyard manure.

COMPOST No. 3.—If 11 maunds of compost No. 1 be mixed with 16 maunds of compost No. 2, the result will be 27 maunds, or 1 ton of compost No. 3.

COMPOST No. 4.—To make compost No. 4, mix 20 seers of bone-dust (made according to the formula given) with compost No. 3, maunds 27, and the result will be 27½ maunds of first-class compost. The proportion of bones may be increased by 10, 15, or 20 seers, if available, but 20 seers of bone-dust is to be the minimum quantity. With increased quantity of bone-dust, an additional quantity of cow's urine should be sprinkled over the compost, and the bone-dust should also be well wetted with cow's urine, and allowed to dry before being mixed with the ton of No. 3 compost. All the bone-dust used should be wetted with cow's urine, sun-dried, and then added to the compost.

NOTE.—According to the tenets of the Hindu religion, the urine of the cow is a sacred and general purifier. The Hindu who loses his caste cannot regain it till certain formalities are gone through, and one of these consists in drinking a prescribed quantity of fresh cow's urine.

If bone manure requires purification in the most effectual manner before being handled, then wetting it thoroughly with cow's urine should effect the object in view. But if the Brahmins, who decide such matters, rule that bone manure cannot be purified by a cow's urine bath, then it is Brahminically demonstrated that this sacred Hindu purificant loses its virtue, when so willed by the Brahmin ; and as the substance—*i.e.*, bone-earth—which deprives it of its sacred and purifying properties is daily consumed by the Brahmin in the milk he drinks and the *Dall* and *Chuppattie* he eats, it stands to reason that if cow's urine cannot purify bone-earth, the Brahmins are hoisted with their own petard.

If, however, cow's urine does purify bone manure, and its contained bone-earth, then the sole objection to its use is at once removed, and all classes of Hindus may turn this valuable manure to account.

112. *Is wood charcoal of value as a manure, and if so give the reason why?*

Yes. Wood charcoal and its grit and dust are of very considerable value as manure, and as components of manures or composts.

Charcoal possesses the property of absorbing and retaining moisture, and when used as a manure and worked into the soil, the moisture invariably present in the charcoal keeps the land from becoming hard and dry. If there be any moisture in the air the charcoal will absorb it. Charcoal also possesses the very remarkable property of absorbing all kinds of gases, and causing them to become latent within its substance or pores.

In this manner 1 cubic inch of charcoal will absorb 90 cubic inches of ammoniacal gas, 35 cubic inches of carbonic acid gas, 9¼ cubic inches of oxygen gas, 7½ cubic inches of nitrogen gas, and 1¾ cubic inches of hydrogen gas.

Charcoal also absorbs large quantities of all deleterious gases, and especially those present in marsh effluvium.

In addition to these properties charcoal is a disinfectant as well as a deodorizer, and when used as a chemical agent induces and produces catalytic chemical action.

113. *Is brick-dust and broken tile and common red pottery dust of any value as a manure?*

Yes. Such dust contains iron in a suitable form for plant food, and if the soil be poor in iron, manuring with brick-dust will supply the deficiency. Most sandy soils are wanting in iron. Many plants, and especially tobacco and poppy, require soil rich in iron; and for these crops brick-dust is a cheap, effective, and valuable manure. If it be desired to increase the proportion of iron, the brick-dust should be wetted with water holding sulphate of iron in solution.

In making saline manures brick-dust is invaluable. It absorbs saline and alkaline solutions, and retains them within its grasp till forced to surrender them to the rootlets of plants in its vicinity.

Brick-dust has a chemical affinity for saltpetre, nitrate of lime, common salt, and sulphate of soda. Nitrate of lime when once absorbed by brick-dust is retained to the last, and if nitrous elements be present in the soil or manure will reproduce nitre by catalytic chemical action. If brick-dust be well wetted with horse urine it becomes rich in urea and ammonia, and all other soluble components present therein. Brick-dust when wetted with cow's urine yields a manure containing the soluble components of cow's urine.

If equal quantities of brick-dust manure so made be mixed together and moistened, after a short time catalytic chemical action will set in ; nitrate of lime will first be formed from its elements, and this, when acted upon by the potash present, will change into saltpetre : and this process will go on till all the potash is converted into the nitrate of potash.

The value of brick-dust as a manure may be considerably increased by adding thereto 5 per cent. of bazar—*Chunam*. Old *Pucca*, mortar reduced to powder, will also answer as well ; but 10 per cent. of it will have to be used in place of 5 per cent. of *Chunam*.

NOTE.—There is not a village which is not freely supplied with its heaps of broken water-pots, chatties, and tiles, as well as broken bricks ; but their manurial value being unknown they remain unused.

114. *What is meant by catalytic chemical action?*

Wood-charcoal and brick-dust induce this peculiar chemical action. They as mediums remain unchanged, but by their agency they cause other substances absorbed by them to undergo a chemical change. For example, if charcoal be immersed for 15 minutes in a strong solution of sulphate of iron, and then taken out and dried in the shade, the pores of the charcoal will show by their iridescence that metallic iron has been deposited within them, and the sulphuric acid has been absorbed. If we take this ferruginous charcoal and immerse it for 15 minutes in a strong solution of saltpetre, take it out and dry in the shade, no peculiar change is perceptible. But if the charcoal be set on fire, the smell of sulphur proves that, though latent, it was derived from the sulphuric acid of the sulphate of iron. The sulphur being consumed, a second odour will become apparent, being due to the evolution and burning of carbon and iron in oxygen gas, which being in excess consumes the carbon thoroughly, and produces an intense heat. When the charcoal is burned out the red colour of the ashes proves that the iron has undergone combustion, and their richness in potash indicates that it was derived from the nitre, which also supplied the oxygen gas.

In reference to the first part of this experiment we see that the charcoal, as a medium of chemical action, remains unchanged, whilst the sulphate of iron has undergone a complete change. In like manner if brick-dust be well wetted with horse urine, which is very rich in ammonia, and other brick-dust (equal quantities of each) be well wetted with cow's urine which is rich in potash, by mixing the two together and leaving the mass alone for some days, chemical action will have set in, and nitrate of potash, or saltpetre, will be formed. Here the brick-dust in each case remains unaltered. But by its means the ammonia of the urine has been converted into nitric acid, which has combined with the potash and produced nitre, or saltpetre.

To show this mode of chemical action more clearly, if we make two aqueous solutions of common salt and saltpetre, and then mix them together and place the vessel on the fire to boil, no change will take place. But if some pieces of charcoal (say 2 to 4 ounces in aggregate weight) be thrown in, and the boiling be continued, chemical change will take place, and when the solution is sufficiently evaporated and left to crystallize, it will be found to contain nitrate of soda and chloride of potash. But the charcoal which produced this chemical change, or catalytic action, will remain unaltered.

NOTE.—"If 16 ounces of charcoal dust or powder and one ounce of sulphate of iron in powder be mixed together, the result will be 17 ounces of a very active and powerful deodorizer and disinfectant. When required for use, place in a dredger and dust freely over the effete matter. Its action is almost immediate. When making use of farmyard manure or turning it over, offensive smells may at once be got rid of by using this simple disinfectant and deodorizer. It acts chemically."—*J. F. Pogson.*

FARMYARD MANURES.

115. *What kinds of animal dung are most commonly employed as manures?*

In England, night-soil, horse-dung, cow-dung, sheep-dung, pig-dung, and bird-dung. In India, in addition to the above, we have elephant-dung, goat-dung, camel-dung, and dog-dung.

116. *Which of these is the most valuable?*

In general bird-dung is most valuable, and is held to include the dung of all kinds of poultry, ducks, and geese. Next to these comes horse, ass, and mule dung; after them, camel-dung, goat-dung, and sheep-dung; then elephant-dung; and lastly, cow and pig dung.

The dung of the dog is extremely rich in bone-earth, but from which a sensible quantity of phosphoric acid has been extracted during the course of digestion.[1]

The dung of the elephant is very rich in woody or cellular fibre, which in time undergoes chemical change, and passes into ***humus***, or humic acid.

117. *Why is horse-dung richer or hotter than cow-dung?*

Because the horse voids little urine as compared with the cow.

Analysis of Horse and Cow Dung.

	HORSE.	COW.	REMARKS.
Potash	9·33	17·15	The Fæces were dried and analyzed by the German chemist Buchner.
Soda	0·61	6·30	
Carbonic Acid	...	0·00	
Lime	5·22	7·31	
Magnesia	2·03	4·50	
Peroxide of Iron ...	2·03	3·34	
Sulphuric Acid	3·92	3·23	
Silicic Acid	59·96	41·00	
Phosphoric Acid ...	7·92	17·05	
	100·00	100·00	

NOTE.—"It appears to me that Liebig's printer has placed the analysis of horse-dung under the head of cow-dung. Inspection shows that, as printed, the cow-dung is made to contain far more mineral matters than horse-dung, whereas just the reverse is known to be the case."—*J. F. Pogson.*

118. *What is the principal objection to using pig-dung?*

[1] It is for this reason the fossil dung of the hyena is deficient in phosphoric acid.

In England, the disagreeable smell and taste it is said to give to the crops raised by the use of it.

In America it is objected to because "it gives a taste even to tobacco plants manured with it."

In India, the lowest class or caste are swineherds, and have nothing to say to the agricultural classes, who object to the use of night-soil as well as pig-dung.

119. *What is the best way of using pig-dung?*

In England and Germany, the best way is to make it into a compost or mix it with the dung of other animals.

NOTE.—Tea, coffee, and indigo planters, who may keep up private piggeries, may at discretion turn the manure to account in the following manner, viz., to one gallon of water add two chittacks of sulphate of iron, and when dissolved bottle for use. When the pigstyes are cleaned out the day's accumulated manure should be sprinkled over with half a bottle of the sulphate of iron water, and then dusted over with charcoal dust, and allowed to dry. The dried manure should then be placed in a pit under shelter, and one chittack of pulverized saltpetre strewn over its surface; the next batch of dried manure to be placed over the first supply, and, as before, sprinkled over with one chittack of saltpetre. This process to be repeated till the pit is full, when straw should be laid on the suface and earth placed thereon and beaten down. A second pit should then be dug, and gradually filled with manure as before and similarly closed, and so on with each succeeding pit.

When the rains are over the manure is to be taken out, mixed with leaf manure and cattle manure, and this mixed manure may be used for all winter grain crops, but not for tea and coffee.

One maund of pig manure to be mixed with five maunds of leaf and cattle manure will yield six maunds of the compound manure to which ten seers of bone-dust and ten seers of *Chunam* or *Kunkur* dust may be added, if it be desired to produce a compost for application to oats, barley, and wheat-barley.

For use as garden manure the bone-dust and *Chunam* may be left out.

120. *Why is cow-dung colder and less liable to ferment than most other kinds of dung?*

"Because the large quantity of urine voided by the cow carries off a greater proportion of that which would otherwise cause it to ferment."—*Johnstone.*

NOTE.—In England, a stall-fed milch cow voids from 2,000 to 3,000 gallons of urine in a year, and this carries with it a large proportion of the soluble saline and other substances derived from the food.

In India, the cow's urine is as a rule wasted.

121. *In what respect does the mixed dung of animals differ from the food on which they live?*

"It differs principally in containing a less proportion of carbon, and a greater proportion of nitrogen and of saline matter than the food they have eaten."—*Johnstone.*

122. *How does it come to contain less carbon?*

"Because animals as they breathe throw off, through their lungs, a large quantity of the carbon of their food." —*Johnstone.*

123. *In what form does the carbon of the food come from the lungs during breathing?*

"In the form of carbonic acid gas."—*Johnstone.*

124. *How much carbon does a man give off, in this form, from his lungs in a day?*

A full-grown man gives off from his lungs about half a pound of carbon in a day, and a cow or a horse eight or ten times as much. "Man gives off more when living on vegetable than on animal food."—*Johnstone.*

125. *Do all the nitrogen and saline matter of the food remain in the mixed dung and urine of animals?*

Yes. "Nearly all the nitrogen and saline matters remain mixed with a smaller quantity of carbon than was contained in the food."—*Johnstone.*

126. *Is this larger proportion of nitrogen and saline matter one cause of the greater activity of the dung of animals?*

Yes; it is one of the principal causes. "Dry matter, in the form of animal droppings, is more fertilizing than an equal weight in the form of the vegetable food on which the animals have lived, for the reason above stated among others."—*Johnstone.*

127. *What form does this nitrogen assume during the fermentation of animal manures?*

"It assumes, for the most part, the form of ammonia." —*Johnstone.*

128. *Is this ammonia a fertilizing substance?*

"It generally is."—*Johnstone.*

129. *How does ammonia enter into the roots of plants when it is formed in the manure?*

"It is dissolved in the soil by water, and is then sucked in by the roots."—*Johnstone.*

130. *What substances are formed in plants by the aid of this ammonia?*

"The gluten and other substances containing nitrogen, are formed in part by the aid of this ammonia."—*Johnstone.*

NOTE.—"It has been shown that one cubic inch of charcoal will absorb 90 cubic inches of gaseous ammonia. Hence it follows that charcoal grit and dust if mixed with cattle dung, as placed in the manure pit, will have the most beneficial effect, and fix the escaping ammonia within its pores. The larger the quantity of ammonia present in the manure, the greater its value as a fertilizer."—*J. F. Pogson.*

131. *Is this ammonia, then, a very important ingredient in our common manures?*

Yes. "Because nitrogen, in some shape or other, is absolutely necessary to the growth of plants."—*Johnstone.*

132. *In which part of the droppings of animals, the solid or the liquid part, is ammonia produced in the greatest abundance?*

"It is produced in the greatest abundance in the liquid part, especially of the droppings of the cow."—*Johnstone.*

NOTE.—"The Hindu zemindar carefully feeds his plough cattle, but cares very little for his milch cows, and less for those out of milk. To show the value of cow's urine, I have freely quoted Professor Johnstone, and trust that at no distant date the cows in every village will be as carefully tended as are plough cattle. The latter are well fed to keep up their strength, and the former should be better fed to ensure rich milk in larger quantity, and also an ample supply of solid and liquid manure of the best quality."—*J. F. Pogson.*

133. *Is it not of great importance, therefore, to preserve this liquid part?*

Yes. "It is of the greatest importance, though it is too often allowed to run to waste."—*Johnstone.*

134. *How should the zemindar proceed in order to make manure from cow-dung?*

It is the custom of the country to convert cow-dung into fuel for culinary purposes, and this manufacture of cow-dung cakes is carried on daily during the spring, summer, and winter months. It ceases *pro tem.* as soon as the rains fairly set in, or during July, August, September in Upper India and the Punjab, and from May to September in Bengal Proper. During the rainy season the cow-dung is wasted and washed away.

To convert it into manure before the rains commence, the village community should unite, and, at some convenient spot outside of the village, construct a thatched shed, open all round. Within this shed a manure pit should be excavated. The pit should be 20 feet long, 12 feet broad, and 4 feet in depth; the shed 24 feet long and 16 feet broad. Height of posts, pillars, or mud walls, to support the thatched roof, 5 feet.

All round the pit a mud wall 30 inches in height should be built to guard against accidents.

The pit being ready, chopped straw, Indian corn stalks, farmyard litter, stable litter, weeds, coarse grass, and leaves of all kinds, should be spread over its bottom to the depth of from 12 to 18 inches.

On this foundation of vegetable matters, the cow-dung, sheep, goat, pony, and horse dung of the entire village is to be daily spread, and as much cow's urine as can be collected daily sprinkled. In addition to this, all the ashes daily produced during cooking operations in the village should also be collected and thrown into the pit, over the cow-dung, or be mixed with it before the carrier starts for the pit.

When the rains cease, a very considerable supply of superior farmyard manure will be ready for use, and this when mixed with the litter, straw, weeds, leaves, &c., now rotted and enriched with cow's urine, will yield a further supply of manure to be mixed with the cow-dung manure; and when so mixed a considerable supply of superior manure will have been secured at the cost of construction of the manure pit, which will last for years.

The soil at the bottom of the pit to a depth of from six to nine inches will be found to be highly impregnated with soluble fertilizing matters.

This enriched soil should be carefully removed for use as manure, and at leisure fresh soil (part of that thrown out when excavating the pit) be laid down in its place.

The manure pit would be village property, and the manure would be the same. The division amongst the village community would be regulated by the number of cows and oxen belonging to each. Thus the zemindar with four cows and four bullocks would be entitled to six shares of manure; two bullocks counting as equivalent to one cow, and so on for larger numbers.

The pit on being emptied in October—November, and laid with fresh soil, becomes available for further use, and may once more be charged with refuse vegetable matters, over which daily supplies of cow's urine, and wood, and cow-dung, fuel ashes, as produced, may be sprinkled, so that a good supply of such manure may be ready for use if wanted before the rains set in, when the past year's process would be repeated.

In large villages several manure pits would be needed, and as by the plan proposed all the cow-dung lost during the rains would be saved and turned into manure, its practical value is manifest.

135. *Is there any important difference between the fermented urine of cattle and the drainings of fermented dung-heaps?*

Yes. "The fermented urine of cows, horses, and sheep contains potash, soda, and ammonia, but no phosphates, while the drainings of the dung-heaps always contain phosphates."—*Johnstone.*

136. *Is there any difference of this kind between human urine and that of the horse, cow, sheep, and pig?*

Yes. "The urine of man and that of the pig contain phosphates; those of the horse, the cow, and the sheep contain none."—*Johnstone.*

NOTE.—"It is necessary to point out from this fact, first, that upon most soils human urine and that of the pig are more valuable as manures; and second, that all the phosphates of their food remain in the solid excrements of the horse, the cow, and the sheep, and also of the dog, goat, camel, and elephant. In Hindoostan human urine is never used as manure; and though in military stations, where European troops are cantoned, large quantities of urine are accumulated every night, the contents of the tubs are carefully removed every morning, taken to a distance, and thrown away.

The zemindars near military cantonments may hereafter utilize this valuable liquid manure, which will keep for some days without becoming putrid if its surface be freely sprinkled over with the charcoal dust and sulphate of iron disinfectant."—*J. F. Pogson.*

BIRD MANURES.

137. *Does bird-dung form a valuable manure?*

Yes. "Pigeon's dung especially is a very rich manure,

and the dung of sea-birds has lately been introduced from South America with great advantage, under the name of guano. When properly applied to land this manure ought to produce at least three times its own value in grain."—*Johnstone.*

NOTE.—The duck and goose yield a manure closely allied to guano. In fact, if these birds were more or less fed on fish, cooked and raw, the manure produced would, if anything, be superior to guano.—*J. F. Pogson.*

138. *Supposing guano to be procurable, how much should be applied per imperial acre?*

"About 2 cwt. per acre, as a top dressing for the grain crops, and 2 or 3 cwt. when used instead of half dung for potatoes and turnips; 1½ cwt. of guano an acre, and 8 bushels of bones or 8 cwt. of superphosphate make a good turnip manure."—*Johnstone.*

NOTE.—"The dung of fowls, turkeys, and pigeons should be daily collected, dried, and stored for use: the object in view is to prevent their rotting."—*J. F. Pogson.*

139. *Which is the most valuable, fresh or rotten dung?*

Weight for weight, rotten dung.

140. *Is it, then, desirable to allow dung to become quite rotten before it is put on the land?*

"By no means; for although rotten dung is more valuable, weight for weight, than fresh, it is difficult to convert the latter into rotten dung without loss."—*Johnstone.*

141. *How do you best prevent loss in fertilizing matter?*

"By carting the manure upon the land, and ploughing it in as soon as possible."—*Johnstone.*

NOTE.—"It is a much-disputed question amongst zemindars and mallees as to whether old or rotted dung should be used in preference to fresh dung. The theory is that fresh dung if applied to the land will burn up the crop sown. In Hindoostan the dung-heap is never under cover, and is exposed to heat, wind, and rain. The consequence is that all gaseous ammonia is expelled by the solar heat, and dispersed by the wind; the rain washes out all soluble fertilizing matters from the dung-heap, and that which remains is the solid and least valuable part of the manure, being composed of the undigested fibres of the hay and grass consumed by the animal as food. This shows how necessary it is to make and preserve manure under cover."—*J. F. Pogson.*

VEGETABLE MANURES.

142. *What is meant by vegetable manures?*

In England, by vegetable manures is meant those parts of plants which are usually buried in the soil for the purpose of making it more productive. In India, indigo refuse, called *Keet;* the squeezed stalks of sugar-cane, called *Megass;* and standing crops of *Moth-dall*, when ploughed into the soil, constitute vegetable manures.

NOTE.—Mr. E. C. Buck, C.S., Secretary to the Government of India, Agricultural Department, in his letter of advice to the Agricultural Society of Bijnor, states: "5. An experiment which appears to promise well is that of green soiling with hemp. Of two parallel plots, both should be ploughed up in July, one allowed to lie fallow, and the other sown with hemp. At the end of August the hemp should be cut down, allowed to rot on the ground, and then be ploughed in with the 'Kaisar' plough. Both plots should then be sown with wheat, and it is believed that the increase of produce on the hemp plot would do far more than counterbalance the cost of the hemp seed and sowing." See Report.

The cultivation of hemp, *Cannabis sativa*, is prohibited in almost all districts, and the native Revenue authorities have full powers to root up and destroy all hemp crops, save those sown for, or on behalf of, the *Abkaree* contractor, or the person who, under the orders of the Department of Excise, is authorized to retail *Gaunjah Bhung*, and *Churrus*, all being obtained from the *Cannabis sativa*.

The zemindars who follow Mr. Buck's advice should obtain a special permit to sow and grow hemp crops by the acre as "green manure;" otherwise they may be heavily fined at the suit of the contractor, and lose their green manure crop into the bargain.

At Kotgurh, as soon as the frosts are over and the warm weather sets in, the first weed that appears is the noxious Indian hemp (*Cannabis sativa*), and the first thing done is to pull the seedling hemp up by the roots,[1] and to collect them in a pit, there to be set on fire when dry and allowed to burn to ashes.

The leaves of the hemp plant contain an essential oil which is injurious to vegetation, and as this oil is not destroyed when absorbed by the soil, it is much more likely to be injurious than beneficial to the growing crop. "A neighbour of mine allowed hemp to grow to the height of two feet in his potato field. It was then dug up and utilized as green manure for the potato crop, which when gathered was found to be very deficient in quantity and greatly deteriorated in size and quality. The hill

[1] Cutting down seedling hemp is of no use, for the roots will throw out fresh shoots; hence the necessity for pulling it up.

mallee said this was due to the *Bhung* being used as manure, in which I fully concurred. The *Bhung* and *Dhut toora* should always be burned, and their ashes used as manure."—*J. F. Pogson.*

143. *In England, by what name are the most important of these vegetable manures called?*

Grass, clover, straw, hay, potato and turnip tops, rape-dust, peat or bog stuff, seaweed, &c.

144. *Are any other plants ploughed in green for the purpose of manuring the soil?*

Yes. "Clover, buckwheat, lupins, white mustard, rape, rye, broom; and in some places even young turnips are ploughed in green to enrich the soil."—*Johnstone.*

145. *Into what kind of soils would you plough in a green crop?*

"Into light and sandy soils, and such as contain little vegetable matter."—*Johnstone.*

NOTE.—"This is exactly what the Indian zemindar does when he allows his crop of *Moth-dall* to come into flower, and then ploughs the entire standing green crop into the ground as green manure."—*J. F. Pogson.*

146. *On the Malabar and Coromandel coasts vast quantities of seaweeds are periodically thrown ashore; are such weeds of any value as green manure?*

Yes; seaweed is a very valuable manure, but its value is not appreciated on the extensive coasts of Hindoostan. But in the United Kingdom, "wherever seaweed or seaware can be obtained in large quantity it is found to enrich the soil very much."—*Johnstone.*

NOTE.—According to Professor Johnstone, 16 loads of seaweed are reckoned equal to 20 tons (270 maunds) of farmyard manure.

147. *How is it employed by the British farmers?*

It is spread over the land, and is either ploughed in or is allowed to rot and sink in, or it is made into a compost. Into the potato drills it is often put in a fresh state, care being taken to prevent the potato sets from touching the seaweed by putting a little earth between them.

"When potato sets are allowed to touch the seaweed they are often observed to rot."—*Johnstone.*

148. *How would you prefer to make a compost of seaweed?*

"I would mix the seaweed with earth and with shell sand or marl, if they were to be had, and would turn over the mixture once or twice before using it."—*Johnstone.*

NOTE.—"*Chunam* or *Kunkur* dust will answer as well as marl, and coral lime or shell lime for shell sand."—*J. F. Pogson.*

149. *Are there any common green vegetables that are ploughed in with advantage?*

Yes; "Potato or turnip tops dug in when the roots are lifted, make the next year's grain crop better."—*Johnstone.*

150. *How can you get the largest quantity of green manure in the form of potato tops?*

"By pulling off the blossoms the tops are kept in a green state till the potatoes are dug up, and thus give much green manure."—*Johnstone.*

NOTE.—"Ample information has been given on the subject of making and using green manure; and it only remains to be added that the leaves of all kinds of culinary vegetables may and should be utilized for this purpose for garden use. Old cabbage stalks, the outside leaves of cabbage and lettuce, beet leaves, carrot tops, artichoke leaves, and the stalks or bines of peas, when green, all yield green manure of superior quality. Never use the leaves of poisonous plants for this purpose, such as *Dhuttoora*, *Nuntoorah*, *Bhung*, or hemp, *vel Cannabis sativa*."—*J. F. Pogson.*

VEGETABLE MANURES MADE FROM SEEDS.

151. *What are rape, mustard, cotton, and lint seed cakes and dust?*

"Cake is the name given to the refuse that remains when rape, mustard, cotton, or lint seed is crushed in the mill to squeeze out the oil. When the cake is broken to powder, it is called dust."—*Johnstone.*

NOTE.—"India, castor-oil cake (*Sessamum*[1] or *Til*-oil cake), and cocoanut-oil cake are available for use as manure. Cotton seed is largely used as cattle food, as well as rape and mustard cake. The exportation of oil seeds is equivalent to the removal of very large quantities of cattle fattening food and valuable manure from Hindoostan. It would be a great gain if all oil seeds were pressed, the cake kept in this country, and the oil only exported to Europe. But as it is, the British purchaser makes his first profit out of the oil obtained by pressing the seeds, and his second profit by selling the oil cake to farmers and rearers of fattened live stock."—*J. F. Pogson.*

152. *How is rape-dust applied as a manure?*

"It is applied to turnips or potatoes, either in the place of the whole or of a part only of the common farmyard dung, and in many parts of England it is used with

[1] "*Til*-oil cake is used as food, and makes confections. It is too valuable to be used as a manure."—*J. F. Pogson.*

great profit, as a top dressing to the young wheat in spring."—*Johnstone*.

NOTE.—"At no distant date India may become a great potato-exporting country, and such manures will be much needed. The ground nut (*Arachis hypogea*) or *Moong phullee*, requires a manure rich in hydrogen and carbon for its vigorous growth, and oil cake supplies the want. It may be laid down as a rule that all oil-producing seeds when sown need a manure rich in the components of oil, and in oil cake this is fully present."—*J. F. Pogson*.

SQUEEZED SUGAR-CANE, OR MEGASS MANURE.

153. ***Can the stalks of sugar-cane, after being passed through the sugar mill, be converted into manure?***

Yes, and into very good manure, and for this reason. Whatever may be the quantity of sugar in 100 lbs. of sugar-cane, one-third of that quantity remains in the squeezed cane or megass, which is used as fuel. Sugar is a natural compound of water and carbon, for 36 lbs. of carbon, and 49½ lbs. of water form 85½ lbs. of cane sugar.

In like manner, the sugar-cane stalk contains in every 81 lbs., 36 lbs. of carbon and 45 lbs. of water.

It has been explained that carbon is of very considerable value as plant food, and analysis proves that it is abundantly present in the squeezed stalks of sugar-cane; therefore to burn it as fuel is waste, whilst to convert it into manure is most advantageous.

In the West Indies it is affirmed that one acre of land will produce 56,673 lbs. weight of sugar-cane, (or 683 maunds). The cane, after being passed through the mill, yields 641 maunds of megass available for fuel.

These 641 maunds of megass when burned will leave 33 maunds of ashes or mineral matters; and if from this quantity (*i.e.*, 1,329 seers) three per cent., or one maund, be deducted for the vegetable alkali or potash, the remainder, or 32 maunds, is the amount of mineral matter drawn out of the soil by the harvested sugar-cane crop.

It therefore stands to reason that the acre of land under sugar-cane has had 33 maunds of mineral matters removed therefrom, and unless the loss is more than made good, the land will yield very poor crops when subsequently sown with cereals.

If the 33 maunds of ashes obtained by burning the megass be applied to the land, the mineral matter will have been restored, but not so the matters rich in carbon; whereas if the megass be preserved and converted into manure, the practical zemindar or European sugar planter secures 292 maunds of carbon, and in addition 33 maunds of mineral matters; or, what comes to the same thing, all the mineral matters present in the megass are extant in the megass manure.

It will be shown in due course that by utilizing the megass in place of using it for fuel, sufficient manure of the best quality will be obtained from one acre of sugar-cane to put six acres of growing cotton under high cultivation; and as this important result will be attained without the aid or use of any farmyard manure, its value will be fully understood by the zemindar and European cotton planter.

TO PREPARE MEGASS MANURE.

With a chaff cutter or chopper, cut or chop the megass as received from the sugar mill, into two-inch pieces, and allow it to dry. Next, under shelter, make a pit 16 feet long, 10 feet broad, and 2 feet deep, and fill it to the depth of six inches with chopped megass, which must now be well wetted or watered with a solution of saltpetre.[1] Then strew prepared pounded *Kunkur* dust over its surface, and then lay down a second course of dry megass, six inches in depth, which water as before, and then a fresh layer of *Kunkur* dust with megass over it, and so proceed till the pit is filled, and has received the last supply of *Kunkur* dust, and been wetted as before with water holding saltpetre in solution.

On the third day the entire surface of the pit is to be well sprinkled with cold water, say four or five "mussucks" full, the object in view being to keep the contents moist, as this produces decay. This process is to be repeated four or five times at the above-stated intervals, *i.e.*, every third day. When the megass has been twenty days in the pit, the contents should be turned over with forks and well mixed together, be moistened for the last time, and be allowed to decompose for another ten days, when the manure should be stored in a dry *Kutcha* well, and beaten down with wood or iron stampers. The mouth of the well to be protected with matting or thatch. The pit being emptied, proceed as before to make a fresh supply.

It is to be understood that the chopped megass in excess of what the pit will hold is to be stored for future use. This should be done near the sugar-works, and if white ants are to be feared, the weakest solution of sulphate of copper (*Neela tooteeah*) dissolved in water, if sprinkled over the drying megass, will protect it most effectually from their ravages.

[1] One ounce of saltpetre to one gallon of water will suffice.

If the weather is fine, the pit need not be under cover; but to be on the safe side, one pit at least should be under cover.

With the soil excavated in making the pit, a walled enclosure, 16 feet long, 10 feet broad, and 3 feet in height, should be made, and used as a surface vat for making megass manure.

All natives understand how to make a mud, or *Kutcha*, wall, and the clay excavated would admit of several mud wall enclosures being made and turned to account as explained. One and all could be thatched over when necessary.

NOTE.—"This formula for making megass into manure was published by the writer in 1870. Shortly after, it was adopted by the sugar planters in the Mauritius, was found to answer expectation, and is now (1882) well known and in general use."—*J. F. Pogson.*

154. *How is Kunkur dust manure prepared?*

Dig any quantity of *Kunkur* which will yield lime if burned, and reduce it to powder in the same way that brick-dust is made from broken bricks and pottery.

Weigh out 32 maunds of *Kunkur* dust and 8 seers of the sulphate of iron (Hindee name, *Heera kusees*). Have ready 8 *Ghurrahs*, or water-pots, filled with cold water, and dissolve 8 ozs., or one quarter of a seer, of sulphate of iron in each, and, having previously levelled and spread out the heap of *Kunkur* dust, proceed to sprinkle the solution of iron over it till the 8 *Ghurrahs* are absorbed; let it dry, and then repeat the dose as before, and so continue till the whole of the sulphate of iron has been dissolved in water and absorbed by the *Kunkur* dust. When dry, dissolve 8 seers of saltpetre in 16

Ghurrahs of water, and sprinkle the solution over the 32 maunds of *Kunkur* dust, and when dry the *Kunkur* dust manure is ready for use for various purposes, and especially for making cotton-plant manure.

155. *How is a compost, or manure, made for cotton plants?*

To every twenty maunds of megass manure, made according to the formula given, add one maund of prepared *Kunkur* dust manure, and mix well together; and in this way prepare as many maunds of cotton-plant manure as may be needed, and store for use if not immediately wanted.

NOTE.—This is essentially a cotton manure, which farmyard manure is not, and the fact is proved by the great production of leaves and wood, but very little cotton wool, from land so manured.

According to Professor Ure, the total constituents of fresh cow-dung are as given beneath.

Analysis of Fresh Cow-dung.

Water	69·58
Woody Fibre	26·39
Animal Matters	2·58
Mineral Matters.	
Common Salt	0·08
Sulphate of Potash	0·05
Sulphate of Lime	0·25
Carbonate of Lime	0·24
Phosphate of Lime	0·46
Carbonate of Iron	0·09
Sand and Waste...	0·28
Total	100·00

The comparative richness of cow-dung in animal matter, and its poorness in carbonate of lime and iron, causes the cotton plant, when manured with it, to run into leaf and wood; hence it follows that, unless the soil is rich in lime, manuring cotton plants with cow-dung only can have but one result, as stated.

In the compost for cotton plants made from megass manure, carbonate of lime is amply present, with potash and iron. But 112 lbs., or 1 cwt., of fresh cow-dung contain only 4 ozs. and 156 grains of lime, and the ton (27 maunds) 5 lbs. 7 ozs. and 61 grains.

Under these circumstances it is simply impossible to expect an increase of cotton wool from cotton plants manured with cow-dung and raised on unsuitable soil.

156. *How do the various mineral matters present in cotton manure act on the cotton plant?*

The carbonate of lime supplies both lime and carbonate to the sap of the plant; iron enriches the sap and makes the plant vigorous and productive; potash causes the lime to dissolve freely, and so prepares it for the roots and rootlets of the plant; and the mineral matters naturally present in the megass provide more iron and all the phosphates.

SUGAR-CANE MANURES.

157. *How is sugar-cane manure to be made?*

The sweet juice of the sugar-cane is a perfect food, inasmuch as it contains an ample supply of carbon in the form of sugar, gluten, and all the mineral matters present in food grains.

To restore the 32 maunds of these mineral matters taken out of the soil by one acre of sugar-cane, a compost must be made of megass manure, bone-dust, *Kunkur* dust, or *Chunam*, mixed with farmyard manure, made under cover, as already explained.

Proportions used in making Sugar-cane Manure.—Take of bone-dust 6 maunds, of prepared *Kunkur* dust, or prepared *Chunam*, 8 maunds, prepared megass manure 20 maunds, and farmyard manure 20 maunds, making a total of 54 maunds, or 2 tons. Mix all well together, when it will be ready for use. To more than make good the loss sustained by the soil after the sugar-cane crop was entirely removed, 108 maunds per acre should be applied to the land, and lightly ploughed in. If a rattoon crop is desired, then the 54 maunds should be dug into the soil, being applied as close to the young rattoon-cane as can be done without cutting roots; and when this crop is removed the other 54 maunds should be applied to the surface and ploughed in.

NOTE.—"It should be carefully remembered that night-soil should never be used for manuring either sugar-cane, sugar-yielding *Sorghums*, or sugar-beet, because this otherwise valuable manure contains a large percentage of common salt, a very small quantity of which is sufficient to interfere with the production of sweet sap in the cane, whilst its presence in beet juice prevents that sugar from crystallizing."—*J. F. Pogson.*

POPPY MANURE.

158. *How should poppy manure be made?*

The bits of broken pottery and brick existing near every village potter's kiln, and outside the *Kulal khana*, or village liquor shop, should be collected and reduced to powder, *i.e.*, made into *Soorkhee*, which is to be wetted with cold water holding saltpetre in solution, and allowed to dry. When dry, it is next to be similarly treated, or moistened with water holding the sulphate of iron (*Kussees*) in solution, after which it is to be dried and stored for future use.

To three parts, by weight, of this prepared *Soorkhee* add one part of farmyard manure made under cover, or two parts if taken from the exposed dunghill. Mix well together, and after the operation the poppy manure will be fit for use.

The process of manufacturing this manure may be carried on at leisure from April to August, and if kept under shelter in a pit until wanted, it will not only keep good, but be greatly improved in strength and quality. The saltpetre present will act on all lime present in the farmyard manure and produce nitrate of lime, which in its turn will be acted upon by the potash of the manure and yield nitrate of potash, or saltpetre.

This manure may be applied either broadcast over the young poppy crop at weeding-time, and so be dug into the soil, or it may be used like common manure, and be ploughed into the land before sowing.

Proportions of Components.

Soorkhee	Maunds	300	
Shorah, or Saltpetre	"	3	
Heera Kussees, or Sulphate of Iron	"	0.18	Seers.
Farmyard Manure	"	100	
Total Maunds		403.18	Seers.

This quantity is to be divided into six equal parts of sixty-seven maunds nine seers each, and applied in the manner indicated to six acres of land (especially selected by those concerned in the Opium Department), which will now bear a rich crop of poppy for three or four years, and yield opium in greater quantity, and of better or higher quality.

The brick-dust used in this manure is rich in the carbonate of iron, and as *Soorkhee* has a chemical affinity for saltpetre and ammonia, its atoms will always contain a certain supply of these fertilizers, and this may be permanently secured by annually manuring the land with farmyard manure, to which saltpetre dissolved in water has been added, in the proportion of twenty seers of saltpetre to one hundred maunds of farmyard manure. Twenty maunds of such manure to the acre will be ample.

In the Simla hills, including Kotgurh and the Rajaship of Busahir, the poppy is extensively cultivated. The seed is sown during the first half of October. The young plant is vigorous and hardy, and is not affected either by the frost or snow of the winter months, and when the spring comes the plants flower, produce full-sized pods, which when bled yield the celebrated opium of the Himalayas, which contains twice as much morphia (7 per cent. to 3½) as the best opiums of Benares and Patna. This shows conclusively that the poppy is a very hardy winter plant. In the plains of India a frost almost destroys the standing poppy crop, and this is beyond doubt due to the deficiency of the sap or milk in iron, potash, and nitrogen. The manure indicated supplies all.

In the plains, the preponderance of water causes the sap to approach the freezing point, and so kills the growing plant.

NOTE.—"The above formula for making poppy manure was published by the writer in 1869–70.

"In 1871, Mr. Phillips, a landholder in the district of Fyzadad, Oudh, published a pamphlet in which his experiments in poppy culture and opium making were given in detail. The manure in question was tried by him, the result being that the land so manured gave twice the quantity of opium as from the same extent of land differently manured. The opium so obtained was officially declared to be, as regards quality, 'better than best.' The pamphlet was presented to the library of the Agri-Horticultural Society of India, otherwise information would have been given in detail.

"The value of the poppy manure has been fully established. But the past failures of the poppy crop shows that the zemindars have not made use of it, from want of information most probably, which could easily have been supplied by the Opium Department, if so minded, as the formula was officially communicated for the consideration of his Honour the Lieutenant-Governor of Bengal."—*J. F. Pogson.*

TOBACCO MANURE.

159. *How should tobacco manure be made?*

The tobacco is, like opium, a narcotic, but it differs from opium in a very peculiar manner, for its leaves possess the power of converting a portion of the nitrogen absorbed from the air and soil into latent ammonia. If a green tobacco leaf be reduced to a paste in a mortar, it has the usual smell of the leaf. But on the addition of a little quicklime, the smell of carbonate of ammonia becomes apparent, and the chemical proof of its existence may be demonstrated by holding a feather, previously dipped in vinegar or muriatic acid, over the paste, when dense white fumes will appear, showing that ammonia is escaping from it in the form of an invisible gas.

All narcotics, and especially the poppy and tobacco plants, thrive in soils rich in potash and iron, and they require an ample supply of nitrogen for their full and perfect development. Nitrogen is present in ammonia and in nitric acid in the following proportions: viz., 17 lbs. of ammonia are composed of 14 lbs. of nitrogen gas, and 3 lbs. by weight, of hydrogen gas.

In the dry state, 54 lbs. of dry nitric acid are composed of 14 lbs. of nitrogen gas, and 40 lbs. of oxygen gas. When this dry acid is mixed with water, the product is liquid nitric acid. It has already been explained that dry nitric acid is present in the nitrates of lime, soda, and potash.

Fresh human excreta, technically called night-soil, is the best manure for the tobacco plant on account of its richness in ammonia. But it must be enriched with potash, bone-dust, or prepared fossil phosphate of lime; sulphate of iron, to fix the ammonia and deodorize the night-soil; and a sufficient quantity of prepared *Soorkhee.*

The zemindars of the upper provinces of India, from Benares to Peshawur, are so averse to the use of night-soil that it is not at all likely to come into use amongst themselves; but all towns and cities having municipalities can easily secure lands in their vicinity, which if manured with night-soil, as stated beneath, would gradually be prepared for tobacco cultivation, and yield a considerable profit if rented to tobacco cultivators.

PLAN OF PROCEDURE.

The ground having been selected, it should be marked off into plots of one acre each—that is, 22 yards broad and 220 yards long, or 44 yards broad and 110 yards in length. At intervals of six feet, a furrow should be made with the common plough, and this should be converted by manual labour into an open trench, not less than nine inches in depth. These trenches being completed, the land is ready for the reception of night-soil.

The night-soil, removed from the city or town, is to be systematically placed in the open trenches, to the depth of three inches, and the soil on each side of the trench is to be raked in over the night-soil, and a ridge made with the superfluous soil. This process of disposing of the daily supply of night-soil is to be repeated till the entire series of trenches have been utilized, when a second set of trenches should be opened between every two covered ones; hence the new trenches would be three feet apart from the old ones, or first set. When these had been similarly enriched with night-soil, the acre of land would have been fully so manured. A second, third, and fourth acre would be similarly treated, and each acre as manured would be ready for the application of *Kunkur* dust; after which, when weather and soil permit, the land should be ploughed at right angles to the parallel series of manured trenches, so as to spread the manured soil over the land, and a second or third ploughing would complete the object in view thoroughly. When the season for transplanting the tobacco seedlings is at hand, manure each acre of land with prepared

Soorkhee, to which a sufficient quantity of bone-dust, or prepared mineral phosphate of lime has been previously added, and plough it in. Finally, make a compost, of two parts of farmyard manure, made under cover, and one part of sheep, goat, or camel dung, and scatter over the surface of the soil, and then plough it in. The acre of land is now fully and properly manured for tobacco cultivation. The reasons for this very high manuring will be found under the head "Tobacco Cultivation."

Fifteen maunds of prepared *Soorkhee*, 15 maunds of prepared *Kunkur* dust, and 8 maunds of bone-dust will be the quantity needed for each acre of land previously manured with night-soil. The compost to be composed of 20 maunds of farmyard or cow manure, and 10 maunds of sheep, goat, or camel manure.

The tobacco plant is the best friend of the sanitary department, for it will consume an amount of effete animal, vegetable, and mineral matter of which few tobacco smokers have any conception. Thus 100 lbs. of dried tobacco leaf contain, on the average, no less than 25 lbs. weight of mineral matter, derived directly from the manured land.

The tobacco of India is inferior in quality to that of Turkey, Persia, Caubul, and America; this is entirely owing to defective cultivation. Tobacco, like tea and opium, requires capital (for it is an expensive and most exhausting crop), of which the zemindar has little, and the ryot none. But if helped by municipalities in the manner indicated, the cultivators would rent the land manured with night-soil, plough as needed, and add the other manures, and so at a moderate expenditure prepare the ground for the reception of seedling tobacco plants.

INDIGO PLANT MANURE.

160. *Can the green stalks and leaves of indigo, after being taken out of the vats in which they have been soaked in water, be converted into manure?*

Yes. The lands near indigo factories would be greatly improved if the steeped green indigo plant was used as manure, being broken up and ploughed in as soon as taken out of the vats. The land so manured, and subsequently enriched with prepared *Kunkur* dust, bone-dust, and farmyard manure, would produce heavy crops of all kinds, including the sugar-producing *Sorghums.* At present the steeped plant is generally piled into a heap, which heats most rapidly, and passes in charcoal. But if on being taken out of the vat each bundle of plant was dipped into a small vat, holding a solution of the sulphate of soda, or *Kharree neemuck*, it would be deprived of its heating powers, and if kept moist would very soon pass into vegetable manure of high quality; which, when mixed with suitable mineral manures, would yield a compost for potatoes, ground nuts, and *Chufas*, or earth almonds. Thus at a very trifling cost a considerable supply of manure would be obtained from a waste substance.

DRY VEGETABLE MANURE.

161. *How is dry vegetable manure made?*

It is to the interest of the agriculturist to grow certain crops, which, after paying all expenses and yielding a profit, will still leave a large supply of refuse matter which may be turned into manure, or plant food.

Of such plants, the sunflower (*Helianthus annuus*) is

the most valuable, for it yields edible oil, and oil cake, and its leaves, stalks, and large seed vessels are available for conversion into manure.

To Prepare such Manure.—As soon as the seeds are ripe, cut off the calyx which contains the seed, and put them into the sun to dry. After the standing crop is removed, cut down the standing sunflower stalks or crop close to the roots, and remove it to the farm. An acre of land under sunflower will give several cartloads of stalk with leaves attached. The stalks should be cut into two or three inch pieces with a chaff-cutter or cropper (*Phursa*), and be allowed to dry in the open. When the whole crop has been so treated, chop up the seed vessels and add them to the heap. When dry, prepare a sufficient quantity of sulphate of soda solution, made by dissolving one chittack of *Kharree neemuck* in a *Ghurrah* of water. Having prepared sixteen *Ghurrahs* of the solution, cause the chopped stalks to be wetted therewith, and then make a second supply of the solution, or as many more as may be needed to wet the entire heap. Turn over the heap, so that every part of it may be moistened; after which, beat or tread it down, so as to compress the mass. It is to be left alone for four days, when the entire heap should be slightly sprinkled with water, so as to keep it damp and moist. This watering is to be repeated at intervals of four days till decomposition sets in, and when this is completed a large supply of dry vegetable manure will be produced.

One thousand lbs. weight of sunflower stalk contain 20 lbs. of potash, and, in addition, other mineral matters; and if this manure be mixed with farmyard manure, prepared *Kunkur* dust, and bone-dust, the compost may be advantageously used for wheat and all grain crops.

This manure is superior to leaf manure, but inferior to megass manure. Its value may be increased if the oil cake of the sunflower seed be mixed with it.

In the Himalayas, all kinds of ground ferns, nettles, and thistles of all kinds in full growth, should be dried and converted into dry vegetable manure. Ferns are rich in potash, whilst 1,000 lbs. of nettles and thistles contain respectively 25 lbs. and 35 lbs. of potash.

RICE HUSK MANURE.

162. *Can the husk of rice be converted into manure; and and if so, how should it be made?*

Every 1,600 maunds of paddy, or unhusked rice, produces at least 600 maunds of husk, which if added to farmyard manure will improve it, and still more so if the rice husk be burned to ashes before being used. The food phosphates do not exist in the grain of husked rice, but they are all present in considerable quantity in its husk; and when this is burned to ashes, the ash left, in addition to a very large proportion of pure transparent silica, contains the phosphates of soda, potash, lime, magnesia, and iron, this last in considerable quantity. Hence it follows that, when rice husk ash is mixed with farmyard manure, in the proportion of three parts of ash with six or eight parts of farmyard manure, a very cheap and valuable manure or compost is obtained, which may at discretion be applied to lands under wheat, barley, wheat-barley, oats, maize, *Holcus Sorghum* (*Jowar*), and the Imperial *Sorghum*, or

"Branching Doorra," just (May, 1882) introduced from Massachusetts, United States of America.

The Chinese are a rice-eating nation, and rarely use milk or milk cheese, but their remarkably intelligent ancestors have taught them how to *manure their blood* (*i.e.*, keeping up a constant supply of the blood, or food phosphates), by boiling their rice in a clear solution of rice husk ash water, and the strength, bone, and muscle of the Chinese shows what the food phosphates will effect.

In Hindoostan, over sixty millions of human beings, who constitute the population of Bengal proper, consume rice daily as their principal food.

The Bengalee boils his rice in plain water, and his physical inferiority proves that the use of food deficient in the food phosphates produces degeneration.

Table showing the composition of the ash of the grain of wheat, oats, barley, rye, Indian corn, and beans, of the straw of wheat, of the turnip root, and of the potato tuber.

	WHEAT.	OATS.	BARLEY.	RYE.	INDIAN CORN.	WHEAT STRAW.	BEANS.	TURNIP BULB.	POTATO TUBER.
Potash and Soda	31	26	32	33	32½	11	45	51½	63
Lime	3	6	2½	5	1½	7	8½	11½	2
Magnesia	12	10	8½	10½	16	2	6½	3	5
Oxide of Iron ...	1	½	½	1½	¼	1	½	¾	½
Phosphoric Acid	46	44	26	48½	45	5	33	11¼	18
Sulphuric Acid	...	10½	2½	1	3	1	4½	15	4
Chlorine	6	¼	5	...	¼	7	1¼	5½	6
Silica	1	2¾	23	½	1½	66	¾	2	1½
	100	100	100	100	100	100	100	100	100

NOTE.—"This most instructive table is taken from

Professor Johnstone's work on Agricultural Chemistry, and shows what an important part phosphoric acid plays in the composition of the cereals, roots, &c., named. Potash is the next most important constituent, wheat containing 28 lbs. to only 3 of soda; whilst the potato tuber contains 63 lbs. of the two alkalies, the potash largely predominating, there being 57 lbs. of it to only 7 lbs. of soda. Magnesia and lime are of next importance. Wheat and wheat straw contain together 13 lbs. of chlorine, derived from common salt used in the manure.

"The value of saltpetre, or nitrate of potash, as a saline manure will be admitted when it is remembered that $101\frac{3}{10}$ lbs. of saltpetre contain 47·15 lbs. of potash whilst 1,000 lbs. of ashes obtained by burning the common Fumitory (*Fumaria officinalis*) will yield only 79 lbs. of potash. Hence it follows that 187 lbs. of saltpetre, which contain 81½ lbs. of potash, are of far more value than 1,000 lbs. of the ash named.

"The nitric acid of the nitre has a specific manurial value of its own, and this acid does not exist in ashes. The plant named does not grow in Hindoostan, but as it contains, weight for weight, more potash than any other plant ash, it has been brought to notice for the sake of comparison.

"The table also shows the value of magnesia, and as this mineral is largely present in steatite, or soapstone, which is freely met with in the bazars, but is at present not used as component of a manure, the information given in the table will, it is hoped, lead to its use. Ten to twelve seers of soapstone dust applied to an acre will supply the magnesia needed for wheat or Indian corn. But double the quantity should be used to meet the requirements of the plant, for the straw of cereals, the leaves and stalk of Indian corn, all contain magnesia, and any of it in excess would enrich the soil."—*J. F. Pogson.*

163. *Do clovers, rye grass, and lucerne contain all the substances named in the table?*

Yes, and in very considerable quantity, as shown in the table given beneath.

Table showing the quantity and composition of the ash left by a ton (2,240 lbs.) of hay of different kinds.

	ITALIAN RYE GRASS HAY.	RED CLOVER HAY.	WHITE CLOVER HAY.	LUCERNE HAY.
Potash	17	26	24¾	30
Soda	17	3½	10½	13½
Lime	3¾	55½	45½	107½
Magnesia	3	17½	14	7¾
Oxide of Iron	1	1½	3½	⅜
Sulphuric Acid	4	6¼	12½	9
Phosphoric Acid	8¾	10	20	29
Chlorine	2	4	5	6¾
Silica	81½	5	6	7⅛
	138	129¼	141¾	211½

NOTE.—"This table is also taken from Professor Johnstone's work on Agricultural Chemistry, and shows why, even when climate admits, the first three grasses do not succeed in India. Lucerne does grow in the plains, but not being properly manured it seldom seeds, and consequently is defective in quality."—*J. F. Pogson.*

LEAF MANURE.

164. *How is leaf manure made?*

In the Himalayas, during the months of April and

May, the forest lands are covered with fallen leaves. If these leaves be collected and watered with a weak solution of saltpetre water, they will soon turn into leaf mould; and if this in its turn be moistened with a weak solution of sulphate of iron, a compost is obtained which, when mixed with farmyard manure, will answer for all gardening purposes, lime and bone-dust being added at discretion.

A perusal of this chapter will show that India is by no means deficient in manures, the most valuable of which, however, are unattainable by the agriculturists. The saline manures being under the lock and key of the Excise, and the manurial wealth of the Sewallics kept out of use from want of cart roads to remove it as quarried.

NOTE.—"The extract given beneath is taken from the *Delhi Gazette* newspaper of the 10th October, 1882, and shows that, whilst the acute men of the Ottawa Valley fully appreciate the value of fossil phosphates, the far more extensive and valuable phosphates of the Sewallics are looked upon with contempt."—*J. F. Pogson.*

"Much interest is being aroused at the present time by the discovery of extensive phosphate fields in the Ottawa Valley. Companies are being formed to work them, and it is expected that the mining operations will result in a large and lucrative export trade."

CHAPTER IV.

WHEAT CULTIVATION.

The extreme antiquity of wheat.—Its cultivation by Adam proved from Scripture.—The finest variety existed in the Holy Land during the lifetime of Joshua the prophet.—This variety apparently still exists, and is now known as the "Golden Grain, or Mammoth Wheat of Palestine."—Detailed list of other varieties named in order of value.—The composition of the grain of wheat.—Chinese method of wheat cultivation.—Mr. Hallett's method of wheat cultivation.—New method of wheat cultivation for India.—American yield of wheat per acre during 1881.—Indian yield of wheat per acre.—The highest known, twenty-two maunds.—The average of 1881, thirteen maunds.—The wheat-straw plait trade of England.

INTRODUCTORY OBSERVATIONS.

IT has been said that modern man found wheat growing in a wild and inferior state, and by cultivation produced the various superior varieties of wheat to be found in different parts of the so-called Old World. That this assumption cannot be maintained will, we think, be admitted when it is known and borne in mind that Adam, on expulsion from Paradise, was doomed to eat bread, under the malediction that, the ground being cursed for his sake, it would bring forth thorns and thistles; that he should eat of the herbs of the field, and eat in sorrow as long as he lived.

Finally, the doom pronounced was conveyed in the following words:

"In the sweat of thy face shalt thou eat bread till thou return unto the ground; for out of it wast thou taken: for dust thou art, and unto dust shalt thou return" (Gen. iii. 19).

From the text quoted may be drawn the conclusion that the bread Adam was to eat was the product of the grain of wheat which even at that very remote period of antiquity was, though self-sown and self-grown, perfectly fitted for human food; and it has so remained unto this day, being still associated with thorns and thistles, which, wherever and whenever they appear, have to be ruthlessly destroyed to secure a plentiful wheat harvest.

When Joshua crossed the river Jordan, and the children of Israel gradually took possession of Palestine, the land was famed for its yield of wheat. The Hebrews in due season sowed the seed-wheat derived from the standing crop when they became the proprietors of the land, and we are told in Scripture that the Israelites were commanded to cultivate the land for six consecutive years; but not to do so during the seventh year, during which the land was to lie fallow and enjoy its sabbath, or year of rest.

A Divine blessing was promised on the crops and harvest grown during the sixth year, and so productive was to be the harvest of the sixth year that there would be ample food for the seventh, or sabbath year, and for the following year as well. It is quite possible that the Hebrews were enjoined to preserve the seed of this triply productive wheat for sowing during the next period of six years, and so on for every succeeding six years. To preserve seed wheat for six years was a very simple affair when the modern discovery of some wheat found in an Egyptian mummy fully three thousand years old proved that it had not lost its vitality or germinating powers, and when sown gave a crop from which the so-called Egyptian mummy wheat of our day was derived.

The Israelites were at all times of their prosperity able to purchase wheat, if needed for sowing, and unless seed-wheat of the first fifth year had been preserved for sowing during the eighth year, they must have purchased their seed wheat from the grain merchants of Egypt and other countries trading with Palestine.

That any description of wheat should, in consequence of a Divine blessing, be capable of producing a triple crop, has been derided, doubted, and scouted. But truth is stranger than fiction; and when, in 1882, we are supplied with a variety of seed wheat originally obtained from Palestine itself, and which yields as an ordinary crop rather more than three times the quantity of wheat per acre than that obtained from any other kind sown and grown on the same extent of land, the humbled infidel and sceptic must admit that, if an acre of land sown with the mammoth wheat of Palestine produces 64 bushels, whilst from 20 to 22 bushels is the average produce per acre of other varieties cultivated in different parts of the world at present, the event recorded in Scripture is fully verified by the very matter-of-fact experience of American farmers and well-known American seedsmen.

It is stated of this remarkable wheat that by merely altering the time of its sowing in a temperate climate, it at once becomes a spring or a winter wheat, and yields its harvest accordingly. In India it will shortly be sown in October and November as a winter wheat, and next spring its value will be determined. To this superb wheat is given the place of honour in this chapter.

165. *What is the botanical and Indian name of the wheat plant, and how many varieties are known to European farmers?*

The common wheat plant is called *Triticum vulgare* by botanists. The varieties are distinguished by suitable Latin names. Thus *T. Hybernum*, *T. Turgidum*, *T. Spelta*, *T. Monococcon*, *T. Compositum* (Egyptian wheat); *T. Polonicum* (Polish, or forty days wheat), and various others. The Indian or Hindee name is *Gae-hoon;* the Persian, *Gun-doom;* Bengalee, *Gom;* Punjabee, *Kunnuck*. The grain is distinguished by its colour into red and white, and also into hard and soft, as soft red, hard red, soft white, hard white.

In the United Kingdom the British farmer cultivates the following varieties of wheat, viz.: The Bearded, the Dunglass, the Golden Ear, the Velvet Ear, the Egg-shell, the Hedge Wheat, the Essex Dun, the Kentish Yellow, the White and Red Essex, the Mungoswell's, the Burwell Red, the Hunters, and the Georgian.

Of foreign wheats the finest varieties are—1st, the Golden Grain, or Palestine Mammoth Wheat; 2nd, Seville Wheat; 3rd, White Russian Wheat; 4th, French Imperial Spring Wheat; 5th, Champion Wheat; and 6th, Defiance Wheat.

166. *Give a botanical description of the grain or seed of wheat.*

"The seeds are yellowish, oval, gibbous, obtuse at both

ends, about two lines long, glabrous, bark convex, marked with a scar towards the summit, and with a longitudinal furrow, margins oval beneath, base downy." —*O'Shaughnessy.*

NOTE.—"According to Professor O'Shaughnessy, 100 seeds of Indian wheat weighed 45 grains. The writer counted out, as they came to hand, 100 seeds of the Palestine mammoth wheat, and they weighed exactly 86 grains, or not far from double the weight of the former."—*J. F. Pogson.*

167. *What is the grain of wheat composed of?*

The composition of the grain of wheat is thus described by the German chemist, Zenneck:

"The *Triticum monococcon* contains, in its unsifted flour, 16·334 of gluten and vegetable albumen; 64·838 of starch; 11·347 of gum, sugar, and extractive; 7·481 of husks. The sifted flour affords 15·536 of gluten and vegetable albumen; 76·459 of starch; 7·198 of sugar gum, and extractive; 0·807 of husky matter. The *Triticum spelta* contains in 100 parts of the finest flour, 22·5 of a soft humid gluten, mixed with vegetable albumen, 74 of starch, and 3·5 of sugar. The ashes of wheat contain the superphosphates of soda, potash, lime, magnesia, and oxide of iron."

"One hundred parts of the bran of English wheat contain, of water, 14; of gluten, 16; of fat, 4; of starch, 43; of woody fibre, 17; of ash, 6."—*Johnstone.*

"The German chemist, Kekule, analyzed gluten from a starch manufactory. It yielded 1 to 1½ per cent. of ashes, which contained 7·87 per cent. of potash, 2·14 of soda, 17·31 of lime, 12·08 of magnesia, 7·13 peroxide of iron, together 47·13 of bases, with 52·08 of phosphoric acid, 0·69 of sulphuric acid, and 0.09 of chlorine."—*Liebig's "Letters on Chemistry,"* p. 436.

NOTE.—"The bran of wheat contains a very large quantity of nutritive matter, and in place of giving it to horses, cows, ducks, and geese, it should, on being sifted out of the flour, be kept for future use. If the bran sifted out of a pound of flour be put in a pan, and be more than covered with water, and boiled for ten minutes, all soluble matters present in the bran will be held in the gruel. If this be run through a sieve, only the woody fibre will remain. The flour from which the bran was sifted should be kneaded with the gruel; and the bread so obtained, whether it be leavened and baked in an oven, or unleavened and made into cakes called *Chuppatties*, will be far more nourishing than the same flour kneaded with water. Most of the iron exists in the bran, and if after the gruel has been separated the dregs be incinerated, the ashes will be found to contain iron. One hundred lbs. of flour contain from 20 to 25 per cent. of bran, and of this not more than two per cent. consists of indigestible woody fibre. Hence, by following the plan suggested, the largest quantity of food will, as a rule, be obtained from the flour so treated."—*J. F. Pogson.*

168. *The ancient Egyptians were agriculturists in remote ages, at which time the Chinese who traded with them were also proficients in agriculture. Can you state how, in ancient times, the Chinese cultivated wheat?*

Yes. In China, the land having been ploughed is ready for the reception of wheat seedlings, which are carefully transplanted from the seed bed, and each plant is manured with night-soil, as farmyard manure is unknown in China.

According to Mr. Davis, "No Chinese farmer ever sows a seed of corn before it has been soaked in liquid manure diluted with water, and has begun to germinate; and experience has taught him (so he asserts) that this operation not only tends to promote the growth and development of the plant, but also to protect the seed from the insects hidden in the ground."

Mr. Fortune, who explored the tea districts of China, states: "During the summer months, all kinds of vegetable refuse are mixed with turf, straw, grass, peat, weeds, and earth, collected into heaps, and when quite dry set on fire. After several days of slow combustion the entire mass is converted into a kind of black earth. This compost is only employed for the manuring of seeds. When seed-time arrives, one man makes holes in the ground; another follows with the seed, which he places in the holes; and a third adds this black earth. The young seed planted in this manner grows with such extraordinary vigour that it is thereby enabled to push its rootlets through the hard solid soil, and to collect its mineral constituents."

In the "Eckeberg Report to the Academy of Sciences at Stockholm, 1765," we are informed that "The Chinese farmer sows his wheat, after the grains have been soaked in liquid manure, quite close in seed beds, and afterwards transplants it. Occasionally also the soaked grains are immediately sown in the field properly prepared for their reception, at an interval of four inches from each other. The time of transplanting is towards the month of December. In March the seed sends up from seven to nine stalks with ears, but the straw is shorter than with us. I have been told that wheat yields a hundred and twenty fold and more, which amply repays the care and labour bestowed upon it."—(*These Extracts will be found in Baron Von Liebig's " Letters on Modern Agriculture," pp.* 250, 251.)

NOTE.—The Chinese system of agriculture brought to notice has been practised for some four thousand years; and its success must be admitted, when it is known that the wheat lands of China are as fertile as ever. The very remarkable fact of the Chinese having traded with the Egyptians in most ancient times, is proved by[1] "*vessels of Chinese porcelain being found in the Pyramids, of the same shape, and with the same characters of writing on them, as are made in China at the present day."—J. F. Pogson.*

169. *Can you describe Mr. Hallett's system of wheat cultivation?*

Yes. Mr. Hallett proves that the land being properly ploughed and manured, the heaviest crop of wheat was obtained by sowing single seeds, at intervals of one foot, or one square foot of land to each seed sown in its centre. Thus, whilst the seed was greatly economized, the heaviest crop of grain was secured.

Mr. Hallett states: "Two adjacent fields, similar in all respects, were sown with the same seed wheat. In the one case, 6 pecks (or 90 lbs.) per acre were sown, and yielded 54 bushels, or 934,000 ears; in the other case, 4½ pints per acre were used, planting them in single grains of one foot apart, and the yield was 1,002,000 (one million and two thousand) ears, or a larger quantity than was produced at the other side of the hedge from more than twenty-one times the seed employed."

[1] *Vide* Liebig's Letters, p. 253.

In his next experiment Mr. Hallett found that "a field planted with 6 pecks per acre yielded only 54 bushels; and one of inferior soil, planted with one peck, yielded 57 bushels, showing that the extra quantity of seed was worse than thrown away."

In another experiment, made for the purpose of demonstrating the possible powers of reproduction in the wheat plant, when properly treated, Mr. Hallett states: "I have now a field of seven acres planted with the produce of a single grain, planted two years ago. One acre of it, with the produce of a single ear planted one year ago."

In reference to another and very important experiment Mr. Hallett states: "The extent to which economy of seed is possible may be guessed from the statement made in reference to 'Tillering,' or horizontal spreading out of the seed roots of wheat, that the stems produced from a single grain having perfect freedom of growth will, in the spring, while lying flat on the surface, extend over a circle 3 feet in diameter, producing at harvest 50 to 60 ears. Now, as an ear contains sometimes 50 grains or more, the above increase is 2,500 at least." Mr. Hallett obtained 1,000 lbs. of wheat from an acre of unmanured land, 1,690 lbs. from an acre with one sort of manure (not named), and 2,000 lbs. with the right sort of manure. —"*Enigmas of Life*," *by W. R. Greg.*

NOTE.—Mr. Hallett's plan of sowing wheat seeds closely approaches that of the ancient and modern Chinese. But the difference is in favour of the Chinese.

In the *Indian Agriculturist* for April, 1882, a correspondent, who withholds his name, supplies a communication purporting to be a "New Method of Wheat Culture for India."

The author of this essay divides an acre of duly prepared land into 1,320 beds, each measuring 9 feet by 2 feet, or 18 square feet per bed, inside measure.

These beds, being hand-hoed, are ready for sowing, which is to be carried out as follows, viz.: "The seeds sown every third inch apart, both ways, by dropping a couple of seeds in five different places within the radius of 4½ inches; in other words, within a space of a diameter of 1½ inches in a circle. The seeds should never be sown deeper than an inch if quick germination of all the seeds be the object, as it undoubtedly is. After sowing, the seed holes should be filled up with earth and lightly pressed, to prevent too quick escape of moisture from the soil."

The quantity of seed required to sow an acre of land on this plan is 2 maunds and 28 seers, or 3½ bushels and 6 lbs., or 14 pecks and 6 lbs. When these seeds germinate and grow up, each double plant will contain from 15 to 20 stalks, and each ear of wheat will contain as a minimum fifty grains of wheat.

The crop when harvested will yield 2,475 (two thousand four hundred and seventy-five) maunds of 82 lbs. each, or 202,950 lbs., or 3,382½ (three thousand three hundred and eighty-two and a half) bushels of wheat. The author concludes: "No computation of straw seems necessary, but for the calculation of the quantity of manure to be applied, the straw produce might, in round numbers, be set down at 3,000 maunds (246,000 lbs.) per acre."—(For details see *Indian Agriculturist* for April, 1882.)

NOTE.—"There are 4,840 square yards in a acre, and as the yield of wheat is put down at 202,950 lbs., each square yard must have produced 41¾ lbs. of grain!

"It has been announced in the public prints that a local government has ordered this essay to be translated into the vernacular for the instruction of its zeminders and ryots, who will very easily prove that the entire scheme is a delusion.

"The author may have heard of the Chinese system of sowing wheat in seed beds at intervals of 3 or 4 inches, and, not knowing the seedlings were to be transplanted, jumped to the conclusion that they were to come to maturity *in situ*, and then at once proceeded to compose his essay, which, however well meant, is simply a theoretical calculation based on error and want of practical knowledge. If a single grain of seed wheat throws out roots covering 6 inches square of ground, 4 seeds will occupy 1 square foot of land. The essayist, however, desires that 2 seeds should be sown in each dibble, or 32 seeds to the square foot, or 28 seeds more than there is growing room for. The theoretical nature of the new system becomes apparent when a reference is made to Mr. Hallett's experiment, which shows that the greatest yield was obtained from single seeds of wheat sown 12 inches apart."—*J. F. Pogson.*

170. *What is the yield of wheat in various parts of the United States of America?*

In the State of New York the average is 14 bushels, and the maximum about 20. In Michigan the average is 11 bushels, and the maximum 18. In New Brunswick the usual yield is 18 bushels. In Canada West 13, in Ohio 15 bushels. Yet in most of these districts the soil is represented to be of almost inexhaustible richness—virgin soil, in fact. One experiment tried in the State of New York, when only 2 pecks of seed were used, showed a yield at the rate of 80 bushels to the acre, or one hundred and sixty fold.

NOTE.—"In the United Kingdom the average is 26 bushels to the acre, and the maximum 60 bushels. The above information shows that India is capable of producing more wheat per acre than does America, and as Indian field labour is very cheap and abundant, there is nothing to prevent India from more than successfully competing with America and Russia for the wheat trade of Europe and England.

"In India the highest yield of wheat per acre is 22 maunds, or 30 bushels and 4 lbs., and the average of 1881, 13 maunds, or 17¾ bushels and 1 lb. Now as this result has been attained without the use of proper manure, it is reasonable to assume that the yield may be doubled when suitable mineral manures are placed at the zemindar's disposal, or within reach of his purse."—*J. F. Pogson.*

171. *Can you state what variety of wheat yields the beautiful straw which has given to England its straw plait trade?*

Yes. The Red Lammas, or White Chittim wheat, from which the straw is obtained, used to come only from Bedfordshire, the slopes of the Chiltern Hills, Herts, Bucks, Oxon, or Berkshire.

Great care is necessary to cut the wheat before the flag of the straw falls. Fair weather is necessary to secure the best straw before cutting. A machine for splitting the straw is needed.

Rhyme of the makers:

> "Under one and over two,
> Pull it tight, and that will do."
> —*Family Herald*, Jan. 18, 1879, p. 191.

NOTE.—"The introduction of this variety of wheat into India would be most beneficial, and would give the native women, widows, and girls an easy and profitable industry, and lead to the formation of an Indian wheat straw plait export trade."—*J. F. Pogson.*

171½. *Can you state how the disease called rust in wheat may be prevented?*

Yes. Rust may be prevented by wetting the corn with a solution of carbolic acid. For 4 bushels take 4 ozs. of Calvert's No. 5 acid added to 2 gallons of water. Place the wheat on a sheet, or tarpaulin, and water with a watering-pot, using the above; thoroughly turn over with a shovel, and sow next morning.

172. *What should a zemindar do to secure a large supply of superior seed wheat from imported stock?*

He should adopt the Chinese system of cultivation: sow the seed in seed beds, and then transplant seedlings into well-manured ground at intervals of 12 inches, or 1 square foot to each seed.

173. *What should the zemindar do to secure the largest yield of wheat per acre from his land?*

He should sow a smaller quantity of seed wheat per acre than he now does. Thickly-sown seed prevents horizontal root growth (tillering) of the wheat plant; and the plants being crowded throw up long, weak stalks, which in time produce small ears of corn, deficient in grain and defective in quality. This is entirely due to the roots not finding sufficient plant food with the restricted circle of their growth.

It is stated by the growers that each grain of Palestine mammoth wheat will throw up from twenty-five to seventy stalks, and as there must be a proportionate development of root growth the advantage of allowing the plants to have growing room, by sowing less seed wheat of all kinds per acre than is now customary, is apparent.

Mr. Hallett demonstrates that 1 bushel of seed wheat (say 30 seers) gave a yield of 30 bushels, or 21 maunds, 37 seers—a quantity not often harvested in India, yet within the reach of all if the land be properly manured.

NOTE.—"To make 30 seers of seed wheat suffice for thin sowing, it should be mixed with the proper quantity of *Kunkur* grits, or broken tile or pottery grits, and then sown broadcast.

"If 2½ bushels, or 1 maund and 34 seers, of seed wheat be the usual quantity per acre, then by the suggested plan 1 maund and 4 seers of grit should be well mixed with 30 seers of seed wheat, and the mixture, on being sown broadcast, the thin sowing desired would be secured, and, as a natural consequence, the crowding of plants be guarded against."—*J. F. Pogson.*

174. *How should the zemindar prepare his seed wheat for sowing?*

By adopting a modification of the Chinese system, and soaking the seed wheat for a quarter of an hour in the urine of the cow. On the day fixed for sowing, the wheat should be placed in a *Ghurrah* and cow's urine poured over it, and poured off as soon as the quarter-hour was up. The grain should then be turned out on a mat and be mixed with the proper quantity of grit and be sown without delay. If white ants, or other ants, exist in the field or locality, half a chittack of sulphate of copper (*Neela-too-teeah*) should be dissolved in each *Ghurrah*-full of cow's urine, and this prepared urine should be poured over the seed wheat as wanted. The white ants will not touch grain which has absorbed the slightest quantity of sulphate of copper; field mice and common ants will also avoid seed so protected.

This system of pickling is applicable to all kinds of grain seeds. It hastens germination and induces vigorous growth *ab initio.*

NOTE.—"The Hindu will not touch liquid manure made on the Chinese plan, or stale human urine on the European plan; but no objection exists to the urine of the cow, which is looked on as sacred by all castes, from the Brahmin down to the Koelee."—*J. F. Pogson.*

175. *The British farmer possesses a machine for sowing corn broadcast upon the ground; another called the drill sowing machine; and a third named the dibbling machine: which of these three would it be advantageous to introduce into India?*

The dibbling machine; as it is established beyond cavil that the most economical and profitable way of sowing seeds of grain crops is by dibbling. The hand dibbling machine would suit India best.

NOTE.—"Broadcast sowing by hand requires the most seed, dibbling the least, and drilling a medium quantity. Broadcast sowing with the machine requires less seed and scatters it more regularly than broadcast sowing by hand."—*H. Stephens.*

176. *Many years ago Mr. Hallett introduced his system*

of pedigreeing wheat, peas, onions, &c.: can you explain how this was done?

Yes. Out of an ear of superior corn the largest and heaviest grain was selected, and this was sown in properly manured land. When the plant came to maturity, out of all the ears produced, the largest grain or seed was again selected and again sown in due season; and this process, extending over some years, was repeated until perfection was attained, *i.e.*, no further improvement in size and quality could take place. Finally, this seed was sown, and all its produce was called "pedigree wheat," which possessed the property when sown of producing at maturity wheat of like quality. Peas, onions, and beans similarly treated gave like results.

177. *Would it be advisable or desirable for the zemindar to improve his inferior indigenous seed wheat by resorting to this system of pedigreeing?*

No; it would simply be a waste of time and money were he to do so, and for this reason: the most expeditious and economical way of securing superior produce is to be attained by sowing superior seed wheat, and preserving the entire yield for re-sowing next year. Imported seed wheat would, under proper treatment, give extra first-class produce; and seed wheat of superior quality obtained from Central India and the Punjab, on being sown elsewhere, would soon displace the inferior local grain.

But such change of seed, and consequent speedy improvement of wheat, is not likely to take place, as the zemindar has no spare cash, and must accept and sow whatever seed wheat (good, bad, or indifferent, or a mixture of all three) his *Buneeah* may choose to give him.

NOTE.—"Mr. E. C. Buck, C.S., &c., in his letter of advice to the Agricultural Society of Bijnor, North-Western Province, states: 'I would strongly advise your recommending the members of your Society to have the best seeds of all kinds selected on their own estates and sent to you for distribution. They should also pedigree their own seed by sowing a few of the best seeds every year on their own farms. A very little suffices. Mr. Fuller can, I believe, give you information how this is done in England.'—*Bijnor Report*, p. 7.

"I very much fear that twenty years of pedigreeing would fail to convert the very best wheat of Cawnpore and Bijnor into a wheat equal to that of Seville or Palestine for size or productiveness. I much doubt if these superb wheats could be improved by pedigreeing; and I hold that pedigreeing of any Indian wheat is incompatible with good husbandry; for why should the zemindar take all this trouble when it is in his power to purchase these two, and other superior wheats, which have only to be sown to yield him first-class produce?

"The Hebrews in very remote times established themselves in Spain, and it is more than probable that the wheat of Seville was originally brought by them from Palestine. I believe it is second only to the mammoth wheat of Palestine."—*J. F. Pogson.*

178. *As the export wheat trade of India is of very great importance, can you suggest how the zemindars could be supplied with the best and most productive varieties of seed wheat, at the same price that they have now to pay for ordinary, inferior, and bad or refuse wheat, supplied by the Buneeah for sowing?*

Yes. The object in view could be easily attained if a

seed farm was formed and kept up in certain districts north of Meerut; the districts selected being those possessing a fertile soil, facilities for obtaining saline and mineral manures (including fossil phosphate of lime), and ample means of canal irrigation.

All empirical experiments should be barred, and strictly prohibited on these seed farms. The best of all varieties of wheat, barley, maize, or Indian corn, broom corn, imported millets, and sugar-producing *Sorghums*, should be sown, grown, and harvested on them, and the acclimatized seed be made available for use by the zemindars of the district on the most favourable terms.

In the Jullunder and Hoshiarpore districts, already famed for cotton culture and high rate of produce, superior varieties of cotton should also be cultivated, and the seed kept for distribution.

These seed farms should belong to the agricultural society of the district, and should receive from the Government a substantial pecuniary grant in aid annually.

NOTE.—"The success of the Bijnor Agricultural Society shows that titled and untitled Talookdars and zemindars of position fully appreciate the value of all sound and practical improvements in agriculture, and there is little, if any, doubt that the Talookdars and zemindars of other districts would follow suit, if the least encouragement was given them to do so.

"Cold, official praise from the Agricultural Department, accompanied by inquisitorial forms to be filled up, does not constitute (at least in my opinion) the introduction and advancement of an improved system of agriculture. What is wanted is cheap money, in place of empirical advice; superior seeds, no matter where produced, or whence obtained, in place of printed forms, and cheap and abundant saline and mineral manures, in place of Excise laws to prevent their being brought into use.

"The land revenue of British India is twenty-two millions sterling, the entire revenue over sixty millions, yet out of this splendid income, one thousand pounds sterling per annum cannot be spared to pay for seeds of cereals, to be obtained as harvested direct from the growers in America, Palestine, Russia, China, Europe, and Great Britain."—*J. F. Pogson.*

179. *In connection with the proposed seed farms, what form of rewards and prizes would be most valued and appreciated by zemindars who undertook the culture of superior varieties of wheat and other food grains?*

The prizes most highly appreciated would be—1st. A milch cow of superior breed, and its calf; 2nd. Silver medals of honour; 3rd. Crimson and gold puggries of honour; and 4th. Silver bangles of honour.

All prizes should be given by Government, the second, third, and fourth being protected from legal attachment as heirlooms, and to be worn by their inheritors.

For reward, the prizeholder to be protected from all possible loss, by the Government undertaking to purchase, at a remunerative price, as much of the prize grain (wheat, barley, and maize) which the zemindar or prizeholder might be willing to dispose of. The grain so purchased to be sent for distribution in districts south and east of Meerut.

180. *If supplied with superior seed of wheat, maize, and barley, could the zemindar successfully compete in the London market with the American farmer for wheat and maize, and with the British farmer for barley?*

Yes. There is not the least doubt on the subject. Upper India, from Meerut to Peshawur in length, and from the foot of the Himalayas to the rivers Sutledge and Indus in breadth, is a vast wheat, barley, maize, and millet producing country. The population is docile, thrifty, and exceedingly industrious, and very expert as agriculturists. The land produces, as a rule, two grain crops (exclusive of rice), the first being maize and millets of various kinds, on which food grains the agricultural population lives, and the second is the cold weather crop, consisting of wheat, barley, gram or chick pea, immense quantities of all being available for export. The barley, though exclusively used by the Punjab breweries, would not satisfy the British brewers, but on first-class seed barley being imported from the United Kingdom, and sown in place of the indigenous barley, the result or produce would, as a rule equal parental stock.

The American farm and field labourer is well fed and paid high money wages. The zemindar's agricultural labourers thrive on *Sorghum* flour, unleavened cakes, and *Dall*, which the American labourer would not touch. The food is at the zemindar's expense. As regards money payment, eight annas (one shilling at par) per month, may be considered as the highest rate of wages, the lowest being eight annas at each harvest, with some yards of coarse cotton cloth.

A little reflection will show that the American farmer cannot compete with the Indian zemindar, whose food and that of his family and labourers is exactly alike, and does not cost a shilling a head per month.

NOTE.—"The average yield of wheat this year (1881) throughout the United States of America was only 10½ bushels per acre, against $13\frac{1}{10}$ (thirteen and one-tenth) bushels last year, being a decrease of 2·6 bushels per acre, or almost 20 per cent."—*Family Herald*, December 10th, 1881.

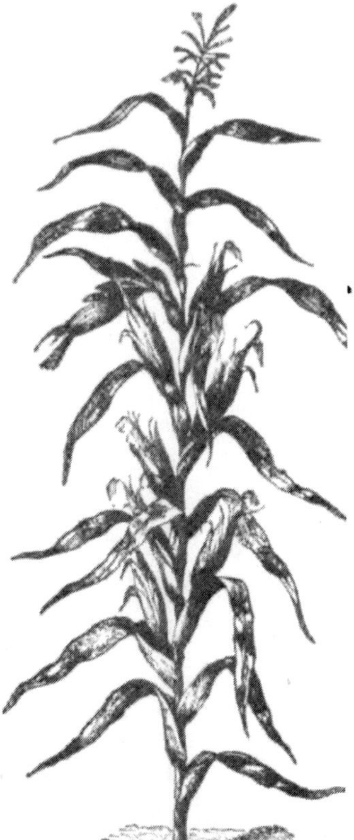

BLUNT'S PROLIFIC FIELD CORN.

"The Indian zemindar, without suitable saline and mineral manures, last year produced; on the average, 13 bushels of wheat per acre. The wheat sown in 1881 and harvested in 1882 gave this yield. With suitable manures this yield should be doubled.

"Without an effort, between February 1881 and February 1882 nearly 900,000 tons of wheat were available for export to Europe, and were shipped from various Indian ports. This quantity may be doubled in 1883, and if the demand is maintained, may in a few years be increased to 9,000,000 tons, if not more."—*J. F. Pogson.*

CHAPTER V.

BARLEY.

Analysis of barley.—Alleged to be indigenous in Thibet.—The wheat-barleys of Little Thibet.—Scotch and French barley.—Pearl barley.—Barley sugar, barley malt, Indian barley.—Its use by Indian brewers of beer.—Parched and given to horses.—Its flour, porridge, and bread or cakes.

181. *What is barley?*

English barley is the ripe seed of the *Hordeum vulgare distichon*, and its ears are two-rowed. The barley of Scotland, called bere, also bigg, is the *Hordeum vulgare hexastichon.* This barley has two rows of ears, but three corns come from the same point, so that it seems to be six-eared. The grains of bere, or bigg, are smaller than those of barley, and the husks are thinner.

The Hindoostanee name for barley is *Jow.*

182. *Can you describe the components of barley flour?*

Yes. 1,000 parts of barley flour contain, according to the German chemist Einhof, 720 of starch, 56 of sugar, 50 of mucilage, 36·6 of gluten, 12·3 of vegetable albumen, 100 of water, 2·5 of phosphate of lime, and 68 of ligneous matter, or husk. The specific gravity of English barley is 1·235. That of bigg, from 1·227 to 1·265. The weight of the husk of bigg is $\frac{2}{9}$.

"The seeds of barley are oval, oblong, alternated at both sides, smooth, straw-coloured, rather angular, with one longitudinal furrow, terminated at the summit by a linear crest, parenchyme white, and farinaceous."—*O'Shaughnessy.*

In the common barley of India, the husk weighs from 18·75 to 25 per cent. The wheat-barleys of Thibet and Kotgurh, in the Himalayas, are devoid of husk.

183. *Can you give the botanical names of the varieties of barley known to botanists?*

Yes. There are fourteen varieties, named as follow: 1. *Hordeum cœleste;* 2. *H. distichon;* 3. *H. distichon nudum;* 4. *H. distichum;* 5. *H. hexastichon;* 6. *H. hexastichon hyburnum;* 7. *H. mundatum;* 8. *H. perlatum.* 9. *H. polystichum vernum;* 10. *H. tetrastichum nudum;* 11. *H. vulgare;* 12. *H. vulgare cœleste;* 13. *H. tetrastichum;* 14. *H. Zeocriton.*

184. *What is the common name for these different varieties?*

1. Is the Celestial Barley of China; 2. The Common Barley of England; 3. Turkish Naked Barley, *i.e.*, has no husk; 4. Battledore Barley; 5. Bigg; 6. Greek Barley; 7. Pearl Barley; 8. Scotch Pearl Barley; 9. Spring Barley; 10. Wheat-Barley; 11. Square Barley; 12. Common Celestial Barley; 13. Bere; 14. Sprat Barley.

NOTE.—"The celestial or Chinese barleys must be identical with those of Thibet, and the Turkish barley no doubt came at some remote period from that quarter. The common or husk barley is indigenous to Thibet, as also are three varieties devoid of husk."—*J. F. Pogson.*

185. *Can you describe these three huskless varieties of*

barley, and give the names by which they are known in the language of Thibet?

Yes. All three varieties resemble the grain of wheat in appearance and form, but have different colours. That most like wheat has also its colour, and its Thibetian name is *No Lona.* The second variety is of a dull pale green colour, and its name is *Tscpa Mar*; and the third variety is brown, the under side of the grain being blackish brown: its Thibetian name is *Nas Mar.* All come under the head of wheat-barley, and should for distinction be called white, dull green, and brown, or black wheat-barley. Their botanical names should be *Hordeum triticospeltum, alba, vert, et nigrum,* all indigenous to Little Thibet and also to Thibet.

NOTE.—"The writer was residing at Kotgurh (at present a missionary station as well as the head-quarters of this sub-district of Simla) during all 1880, and till October, 1881. Two kinds of wheat-barley were found to be growing in Kotgurh, and the oldest inhabitant could not tell whence the seed was originally obtained. Thinking that the seed of both kinds, white and dull green, might have come from Little Thibet, samples of both were sent by post to the Rev. E. Pagel, missionary, Church Missionary Society, with a letter inquiring if these varieties of barley had been cultivated in Little Thibet and at Poo, the locality of the mission station. In due time the reply was received, and with it four ounces of each of the three varieties of wheat-barley, two of which had been cultivated from time immemorial by the zemindars of Kotgurh, whilst the dark brown variety was quite unknown to them. Thus the fact was absolutely established that all three varieties of wheat-barley were indigenous to Poo, in Little Thibet, and throughout the country. Mr. Pagel stated that the common or husk barley was cultivated as food for cattle, whilst all three varieties of wheat-barley were cultivated for use as human food; the flour being used to make unleavened bread or cakes as well as porridge, and the malted grain being made into beer and spirits.

186. *Have any of these Thibetian wheat-barleys been cultivated in the plains of India; and if so with what results?*

Yes, during 1881–82. The white and dull green varieties have been most successfully cultivated. The former on lands belonging to the Agricultural Society of Bijnor, in the North-Western Provinces, which gave a yield of 22 maunds, or 2,804 lbs., to the acre; and the latter by Mr. W. C. Peppe, of Birdpore, Goruckpore. Of it the grower reports: "The wheat-barley promises to be a great success. It was the first ripe and the first cut of any *Rabbee* crop about here. The standing crop was a magnificent sight. The produce from four-fifths of an acre, manured with bone-dust, was 1,156 lbs., and straw, 1,640 lbs. The produce from two-fifths of an acre, manured with ordinary farmyard manure, was 255 lbs. of grain, and of straw 410 lbs."

Mr. R. Nicholson, of Mirzapore, North-Western Provinces, has for some years very successfully cultivated this variety of wheat-barley, of which he was the first grower, having been presented with a handful or so of seed by a servant, to whom it was given by a Hindoo pilgrim, probably from Nepaul. During 1880–81 this barley was cultivated in the Horticultural Society's Garden at Lucknow, and gave a yield of between 22 and 23 maunds of grain per acre. During the sowing season of 1881, Mr. Louis Dane, of Her Majesty's Bengal Civil Service, caused one maund of mixed white and dull green wheat-barley seed, grown by the zemindars of Kotgurh,[1] to be sown, partly on land attached to the Government House at Lahore, and the remainder on land belonging to the Horticultural Society, Lahore. The results are given below.[2] Mr. A. Shircore, of Jullunder, grew both varieties successfully. The black wheat-barley has not as yet been cultivated in the plains.

NOTE.—"About two maunds of the mixed wheat-barley, value 6s. 8d., or four rupees, was sent by order to Mr. Buck, head of the Agricultural Department, on the same day that Mr. Dane's supply of one maund was despatched. The seed so sent did not reach Cawnpore from Simla in time for sowing during the season of 1881. This unaccountable delay is much to be regretted, as the writer

[1] Kotgurh is 50 miles north-east of Simla. Its altitude is 6,400 feet above the sea level. After the 10th November there is frost every night; the snow falls in December, January, and February. It vanishes in March, when the barley and wheat crops commence spring growth.—*J. F. Pogson.*

[2] Mr. Edgar Spooner, Superintendent of the Agri-Horticultural Society's Gardens, Lahore, reports that $\frac{1}{16}$ of an acre of land manured with nightsoil at the rate of 200 maunds to the acre, and sown with 5 lbs. of wheat-barley, gave of grain at the rate of 27 maunds and 24 seers per acre, and of straw 39 maunds. The same quantity of seed sown on the same extent of unmanured land gave of grain at the rate of 19 maunds 32 seers per acre, and of straw 34½ maunds. The difference in grain — 7 maunds and 32 seers per acre—shows that nightsoil as manure should not be applied to land under wheat-barley cultivation. If the land produced over 19½ maunds without manure, it should have produced just double with the proper description of manure.—*J. F. Pogson.*

gave Mr. Guthrie, of Dehra Doon, North-Western Provinces, some of the mixed wheat-barley seed on the 20th November, 1881, and which was sown three or four days after. It germinated perfectly, and though sown so very late in the season it was nevertheless the first barley to ripen seed; the ordinary barley of Dehra Doon growing alongside did not ripen seed for a week after. The produce was very satisfactory. These results show conclusively that wheat-barley suits the soil and climate of Upper India admirably."—*J. F. Pogson.*

187. *Will common barley or wheat-barley grow and ripen seed in the Presidencies of Madras and Bombay?*

No. Though the seeds will germinate, the plants will not come to maturity on account of there being no winter properly so called in those parts of Hindoostan. In Madras the maize crop is harvested in December, which is only a winter month in name. In Bombay iced wines and punkas are indispensable at Christmas, and after it. Hence no winter exists; and as barley requires for its growth from four to five months of genuine winter, with frost in the plains and frost and snow in the Himalayas, it is obvious that it will not come to maturity in any part of Hindoostan where the winter months do not possess the corresponding temperature.

NOTE.—"It has been necessary to explain this matter fully; for though well known to all boys attending agricultural schools in England and Scotland, the fact of barley requiring a *bonâ fide* winter for its production is not known to high Indian officials. In September and October, 1881, the head of the Agricultural Department ordered the writer to send a supply of seed wheat-barley to the Model Farm at Madras, and also to Bombay:

and as Mr. Buck did not seem to know that barley would not grow at the farms mentioned for the reasons given, he was informed accordingly. On the 19th of July, 1882, a Madras official, in charge of a large estate belonging to a minor, had to be informed that his request to be supplied with wheat-barley could not be complied with, as the grain would not grow in Madras.

"The reader should bear in mind that to bone manure Mr. Peppe ascribes his success in wheat-barley culture." —*J. F. Pogson.*

188. *In what part of Hindoostan is the best husked barley produced?*

In the Principality of Reewa, in Rajpootana.

NOTE.—"The introduction of this barley into the North-Western Provinces is much to be desired, and may shortly be taken in hand by the energetic secretary of the Bijnor Agricultural Society. Had Government seed farms existed, such a measure would not have been necessary."—*J. F. Pogson.*

189. *Is it desirable that wheat-barley should be largely cultivated in Upper India and the Punjab?*

Yes, it is most desirable; and once introduced would become very popular and lead to the cultivation of the common husked barley being abandoned. It is said that 100 lbs. of ordinary Indian husked barley contain 25 lbs. weight of husk; and as the wheat-barley is devoid of husk, the farmer or zemindar would be a considerable gainer by growing it, and secure an ample supply of food for his household, and yet have enough for seed and for sale.

NOTE.—"The Indian brewers would purchase wheat-barley largely for conversion into malt, and in time very considerable quantities would be exported to Europe and England for brewing and distillation into very superior whiskey. Indian whiskey distilleries would be called into existence, and its export to Australia and elsewhere would soon follow. Such whiskey would be free of the noxious fusel-oil, and, as a rule, command the highest prices.

"To show the Talookdars and zemindars of Oudh, the North-Western, Provinces, and the Punjab what an immense demand there is in England for superior barley, it is only necessary to state that 'Mr. Bass's brewery at Burton-on-Trent requires the product of seventy thousand (70,000) acres to furnish barley enough to run the works for one year.' There are numerous other brewers, but none who consume barley so largely. The German brewers also work and brew beer on the largest scale; and as India can supply unlimited quantities of barley, an immense export trade in barley may be secured as soon as the three varieties of Thibetian wheat-barley are extensively cultivated. The quantity of barley distilled is also very great, and is, in addition to that, used by the brewers of England, Scotland, and Ireland."—*J. F. Pogson.*

190. *Can you state which of the varieties of barley cultivated by the farmers of England and Scotland might with advantage be introduced into Upper India?*

Yes. All the varieties of barley which, under suitable treatment, are capable of being manufactured into pearl barley, should be introduced, with the object of calling this industry into existence, and so providing the agricultural population with a (to them) new article of very wholesome and valuable food. In addition to this, large quantities of pearl barley would be manufactured for export, and the producer would be directly benefited.

191. *Can you explain what is meant by barley sugar?*

Yes. Pearl barley boiled in a sufficient quantity of water yields a decoction which is run through a sieve to separate the grains of boiled barley. To this decoction, when cold, the proper quantity of sugar is added, and the syrup is boiled down to viscidity, run into a plate, cut into strips, dried, and preserved for use. The sweetmeat so made is called barley sugar. Patent barley is pearl barley ground into flour, and barley water is gruel made with barley flour.

192. *Why should not the other varieties of barley cultivated by Continental and British farmers be introduced into India?*

For the following reasons. 1st. All are husked barleys and if grown in India any increase to the recognized weight of husk would immediately lead to a sensible reduction in price. 2nd. The freight per ton of wheat-barley devoid of husk, and first-class barley with husk, is identical. But as five tons of wheat-barley and six tons of common barley equal each other, the merchant who exports six tons of common barley pays for one ton of *valueless husk,* whilst the one who exports five tons of wheat-barley has no husk to pay for. 3rd. As all the so-called pearl barleys are provided with husk, and are also considered the best of such husk varieties, their culture for all purposes is to be preferred to other varieties which may in a moment be condemned as having too much husk.

193. *What description of soil is best suited to the culture of barley?*

Barley requires a clean, rich, mellow loam, moderately retentive of moisture. Clay soils tempered with sandy mould also suit barley; soils containing a due proportion of lime, clay, and sand are likewise adapted to barley culture. But on poor, wet soils it is never successful. The land to be sown with barley should be well ploughed and thoroughly pulverized. Barley should be sown as soon after ploughing as possible, when the land is fresh and moist, in order to obtain equal and speedy germination.

194. *What manure is especially needed by barley crops?*

Prepared *Kunkur* dust, mixed with bone manure, or fossil phosphate of lime; and when about six inches in height, a top dressing of prepared farmyard manure should be applied. This will not only benefit the young crop, but prevent hares, antelopes, and all kinds of deer from eating the same, which they will do with impunity if the growing crop is not so protected.

195. *What quantity of seed barley is needed to sow one acre of ground?*

The reduced quantity recommended for wheat in the preceding chapter. All barley, and especially wheat-barley, is given to tillering, or horizontal root growth, and if there is not growing room for this purpose the plant throws up long thin stalks with only two or three ears; whereas with sufficient room for tillering ten to fifteen ears and more are formed on as many stalks, all being the produce of a single seed. Such stalks are thick, the leaves deep green, broad and fully developed, whilst the growth of the entire plant is most vigorous.

NOTE.—"Mr. Pagel informed the writer that more than twenty years ago one of the three varieties of wheat-

This shows that if, at some future time, cargoes of all three kinds of wheat-barley are sent to Trieste, they would meet with ready and profitable sale, owing to one variety being known and grown in Germany. The British distillers of whiskey would be only too glad to purchase any quantity of wheat-barley raised in the Punjab; but, unfortunately, the Punjab does not possess a single public or private seed farm whereon seed barley, obtained from Kotgurh and Poo, could be grown under high culture, and the acclimatized seed be preserved for distribution."—*J. F. Pogson.*

196. *Is barley much used in India as horse food?*

Yes. Parched with equal quantities of *Gram* it is the daily food given to horses, and very large quantities are used in the studs and cavalry regiments. The common or husk barley is so used.

Table showing the composition of 100 lbs. of Barley Ash, obtained by careful incineration of a sufficient quantity of Barley.

Potash and Soda	32
Lime	2½
Magnesia	8½
Oxide of Iron	½
Phosphoric Acid	26
Sulphuric Acid	2½
Chlorine	5
Silica	23
	100

NOTE.—"One hundred lbs. of barley when burned will yield only three lbs. of ash, and in it potash and phosphoric acid are the principal mineral matters. Hence the necessity for bone manure or fossil phosphate of lime and potash."—*J. F. Pogson.*

CHAPTER VI.

OATS.

Botanical and Indian names.—Analysis of the grain.—Varieties of oats under cultivation.—Mode of cultivation.—Used as horse food.—Oatmeal used as human food, as also groats.—Analysis of the ashes of the oat.—Oat cultivation in the Bengal Presidency.

197. *Can you explain what is meant by the word oats?*

Yes. Oats constitute the ripened seed of the *Avena sativa*, or cultivated oat plant. The natives of Hindoostan call it *Jai*, in contradistinction to barley, which is called *Jow*. According to botanists, the native country of the oat is unknown; but modern discoveries point towards China as the home of the oat, and it may also be indigenous to Thibet. Wild oats are common in the Kotgurh hills, and in Busahir, whilst the cultivated oat is unknown.

198. *Can you give a description of the grain and of its components?*

Yes. The seeds of the oat are cylindrical, unguiculate, covered with a black or white envelope, smooth, furrowed, or angular; grain naked, villous, pale grey, bearded at the summit; parenchyme snow-white, farinaceous. Taste sweetish, mucilaginous.

The German chemist, Vogel, found that 100 parts of oats afforded 66 parts of flour or meal, and 34 parts of bran. The flour contains 2 parts of a greenish yellow fat oil; 8·25 of bitterish sweet extractive; 2·5 of gum; 4·30 of a grey substance, more like coagulated albumen than gluten; 59 of starch; and 24 of moisture, inclusive of loss.

The German chemist, Schrader, found in the ashes of oats silica, carbonate of lime, carbonate of magnesia, alumina, with oxides of manganese and iron.

199. *Can you name the varieties of oats under cultivation in different parts of the world?*

Yes. The Chinese cultivate a variety of the oat called Chinese hull-less oats (*Avena nudum*, or naked oat). These thresh directly from the straw without a particle of hull or chaff adhering, the grain weighing as high as 55 lbs. to the measured bushel. It ripens earlier than the common varieties.

THE WHITE OATS OF RUSSIA (*Avena sativa alba*) takes the first place as a hulled oat, one ounce of seed giving 157 lbs. One seed has been known to produce 22 stalks growing from one stool, and the product from one seed was not less than 1,500 grains or seeds. The usual yield of the white Russian oat is 100 bushels per acre. Its introduction into India is much to be desired.

THE WHITE GERMAN OAT, called *White Probsteier Oats*, would also suit India. The yield varies from 56 bushels to 98 bushels, the average being about 74 bushels. The measured bushel of this oat does not exceed 39 lbs. Hence its inferiority to the Chinese oat.

THE WHITE ZEALAND OAT. This variety is highly commended, the straw being very tall and thick, and the leaves remarkably broad. The heads were 15 inches in length and well filled. This variety would thrive in Upper India, but is at present quite unknown.

NOTE.—These four varieties of oats have been perfectly acclimatized in America, and the seeds of all may be procured from Mr. James J. H. Gregory, Marblehead, Massachusetts.

THE SPANISH OAT (*Avena strigosa*), or thistle-pointed oat, would also suit an Indian soil and climate. Probably Central India would be the best locality for its successful cultivation.

The British farmer cultivates three varieties of oats, of which the variety known as the "Hopetown" does not litter out, hence requires more seed to be sown.

The best British oats are raised in Scotland, and all varieties which thrive there should be introduced into Northern Hindoostan, including the Punjab.

The oat is suited to climates which are too cold for wheat and barley. Hence its introduction into the inner and superior ranges of the Himalayas would be a boon to the population, who leave all ground which will not ripen barley uncultivated. The varieties of British oats which tiller, should, as a rule, be preferred to varieties which do not tiller.

NOTE.—Messrs. Sutton & Sons, Reading, Seedsmen to Her Majesty the Queen, Empress of India, will on indent supply all varieties of oats cultivated in Scotland.

200. *What quantity of British oats should be sown on an acre of prepared land?*

The quantity of seed necessary varies from four to seven bushels per acre, and broadcast sowing is that generally practised. If the soil and climate be favourable, the maximum yield may be estimated at 70 bushels, and the minimum at 20 bushels per acre. Oat straw is preferred to any other as fodder for cattle, as it is considered more nutritive.

201. *What soil and manure does the oat crop require?*

Soils suited to wheat and barley will suit oats as well, and similar manures should be used. The oat, like barley, needs phosphate of lime, either as present in prepared bone manure, or in the fossil phosphate of lime, so abundantly present in the Sewallics.

NOTE.—"In the United Kingdom, a change of seed oats from hot climate to cold, and cold to hot, is always to be recommended. Hence the immense value of the Inner Himalayas, from Kotgurh to Murree, as localities for acclimatizing superior varieties of wheat, barley, oats, rye, and all kinds of maize. The success of the Reverend Mr. Carleton, missionary in Kulu, shows what may be accomplished by a seed farm, even on a small scale, and it is to be hoped that sooner or later his example will be followed by the Government concerned. In no country but India would a poor missionary be saddled with the cost and trouble of undertaking and satisfactorily carrying out agricultural improvements, from which the population benefits in the first instance, and the State as soon as they are thoroughly established."—*J. F. Pogson.*

202. *Would the agricultural population as well as the Government be benefited if oat cultivation was extended and fostered?*

Yes. When the people of Upper India come to understand the value of oats and the resulting oatmeal and groats as an article of food, the cultivation would become very popular. In the Punjab, where horse and mule breeding is steadily increasing, the production of oats on the large scale for use as horse food is a desideratum. The Government studs and the regiments of cavalry and horsed batteries of artillery could be fed on oats, and the zemindars who supplied them would be considerably benefited.

203. *Can you explain what is meant by the word groats?*

Yes. Groats, or cutlings, are the bruised oat seeds freed of the pericarp. They are much used for preparing gruel for the sick. Porridge, a valuable article of diet, is made by stirring groats or oatmeal into boiling water until the mixture thickens. Oatmeal cakes are most nourishing, oatmeal being richer than any other kind of meal, both in gluten or nitrogenous matter, and in fat.

204. *Of the three cereals, wheat, barley, and oats, can you state which yields the most abundant supply of straw, and which is the most nourishing?*

Yes. Under high cultivation in England, an acre of land under wheat has produced 3,476 lbs. weight of straw, barley about as much, and oats in America have produced more straw; this is owing to the superior varieties of oat, possessing straw five feet in length, and of sufficient thickness and strength to stand erect with its load of produce. Oat straw is the most nourishing, wheat and barley straw being about equal.

NOTE.—"When it is borne in mind that the superior varieties of oats brought to notice yield 100 bushels of threshed oats per acre, and an abundant supply of the most nourishing straw as well, the importance of the extension of oat cultivation will be generally admitted by all sound thinkers."—*J. F. Pogson.*

205. *What varieties of oats are cultivated in Bengal?*

When the studs were established, many years ago, the common oat was introduced, and has since been cultivated at all stations having Government studs. The seed never having been changed has degenerated. The superior varieties are quite unknown to the agricultural authorities of India.

Table showing the composition of 100 lbs. Ash, obtained by incinerating a sufficient quantity of Oats.

		REMARKS.
Potash and Soda	26	Barley ash contains 23 per cent. of silica and 2½ of sulphuric acid. Hence for horse food the oat is superior to husk barley.
Lime	6	
Magnesia	10	
Oxide of Iron	½	
Phosphoric Acid	44	
Sulphuric Acid	10½	
Chlorine	¼	
Silica	2¾	
	100	

WHITE ZEALAND OATS.

CHAPTER VII.

RYE.

Botanical name.—Has no Indian name.—Analysis and description of the grain.—Varieties of rye under cultivation.—Mode of cultivation.—Used as a breadstuff.—Malted and distilled into rye spirit, called Gin.—Rye cultivation in the Punjab advocated.—Analysis of the ashes of rye.

206. *Can you explain what is meant by the word rye?*

Yes. The ripe seeds of the rye plant (a cereal) are so called. Its botanical name is *Secale cereale.* There are two varieties, *S. cereale Hybernum*, or winter rye; and *S. cereale vernum*, or spring rye. Rye is quite unknown in the plains of India, and has no Indian or Hindee name.

NOTE.—"The Moravian missionaries at Keelong, in Lahoul, on the frontier of Thibet, have introduced and successfully cultivated rye on the mission lands. The climate of the locality is severe in winter. Hence it follows that rye will thrive in all places having a suitable soil and milder winter. Rye would grow throughout the Himalayas north of Simla."—*J. F. Pogson.*

207. *Can you give a description of the grain and of its components?*

Yes. The seeds of rye are grey, conical, two lines long, acute, glabrous, back of seed convex, marked with a longitudinal furrow, base obtuse, keeled and oval scar towards the point, surface slightly folded. The flour of rye is white, and the bread made therefrom is nourishing and very wholesome. The rye plant is said to be a native of the borders of the Caspian Sea.

According to the German chemist, Einhof, 100 parts of rye meal contain 24·2 parts of husk, 65·6 of flour, and 10·2 of water. This chemist found in 100 parts of the flour 61·07 of starch, 9·48 of gluten, 3·28 of vegetable albumen, 3·28 of uncrystallizable sugar, 11·09 of gum, 6·38 of vegetable fibre, and the loss was 5·62 including a vegetable acid. Some phosphate of lime and magnesia was also present.

208. *What quantity of seed rye should be sown on an acre of prepared land?*

The British and American farmers allow one and a half bushels of seed rye to the acre. The bushel of rye weighs 56 lbs.; hence 84 lbs., or one maund and one seer, per acre would be needed by the zemindar who undertakes the cultivation of this hitherto unknown but valuable cereal.

209. *What soil and manure does the rye crop require?*

"In the United Kingdom rye is usually sown on light lands, and does not require so much care as wheat; it suffers less by being sown on the stubble of another corn crop, or upon its own, and it is not unusual to grow it on the same land two years in succession. This grain is frequently sown to be cut for soiling instead of winter tares, and in England it is frequently used for early sheep feeding, cut green, without obtaining a grain crop from it. Rye is extremely useful to breeding flocks, as it comes forward earlier than tares, and affords good food when other sustenance is scarce."—*Vide* "*Agriculture*," *Chambers's* "*Information for the People*."

The manure needed for barley should also be applied to land to be sown with rye. Bone manure or fossil phosphate of lime should never be omitted.

NOTE.—"Rye not having as yet been cultivated by the Cawnpore Agricultural Department, or by myself, I have been obliged to make the above extract. To my good friends the Moravian missionaries belongs the credit of having obtained seed rye from Germany, and introduced this valuable cereal into the Thibetian Himalayas. The Punjab has a prolonged but mild winter, and a very considerable extent of arable lands suited to rye cultivation. But the Punjab, which needs it most, does not possess an Agricultural Department, and it is not at all likely that the well-to-do but uninformed zemindars of the Punjab will enter on agricultural improvements for the ultimate benefit of the ruling powers, who should take the lead in such matters, and import and acclimatize all superior varieties of cereals in localities possessing suitable soil and climate, and not leave this most important work to philanthropic Europeans and poorly-paid missionaries."—*J. F. Pogson.*

210. *Would the agricultural population as well as the Government be benefited if the cultivation of rye was introduced and fostered by the officers of Government in charge of districts?*

Yes. Both parties would be benefited. The zemindar would obtain a new and valuable food grain, easily raised, and, if produced in sufficient quantity, capable of being exported to Europe at remunerative rates. The Government would be directly benefited through the

Department of Excise. India at present imports every quart of gin drank in the country. Gin cannot be made or distilled without rye, hence at present India cannot produce this popular spirit for local consumption as well as export, and until private enterprise enters the field, and gin distilleries are established, especially in the Punjab, no increase of Excise duty or revenue can be expected.

NOTE.—"The materials employed in the distilleries of Schiedam are two parts of unmalted rye from Riga, weighing about 54 lbs. to the bushel, and one part of malted barley (bigg), weighing about 37 lbs. per bushel. The quantity of spirit (gin) varies from 18 to 21 gallons per quarter of grain; this large product being partly due to the employment of the spent wash of the preceding fermentation, an addition which contributes at the same time to improve the flavour."—*Vide Dr. Ure's Dictionary*, "*Gin*."

211. *Can you explain how rye bread is made, and how it differs from wheat bread?*

Yes. Rye meal after being sifted is made into dough in the usual manner, yeast is added, and the leavened dough when baked yields a true spongy bread, which is reckoned very wholesome. Wheat bread is white, whilst rye bread is dark-coloured. Rye bread is very extensively used in France, Germany, Belgium, Denmark, and Russia; hence all these countries would largely purchase rye grown in the Punjab.

Professor Johnstone states: "In composition and nutritive quality wheaten and rye bread very closely resemble each other; and except as concerns our taste, it is a matter of indifference whether we live on the one or the other. Rye bread possesses one quality which is in some respects a valuable one; it retains its freshness and moisture for a longer time than wheaten bread, and and can be kept for months without becoming hard, dry, or unpalatable."—*Vide Johnstone's "Chemistry of Common Life.*"

Composition of the Ash of the Grain of Rye.

		REMARKS.
Potash and Soda	33	Rye ash contains more phosphoric acid than the ash of any other cereal. Hence the necessity for phosphate of lime as a mineral manure.
Lime	5	
Magnesia	10½	
Oxide of Iron	1½	
Phosphoric Acid	48½	
Sulphuric Acid	1	
Chlorine	0	
Silica	½	
	100	

CHAPTER VIII.

RICE.

Rice, its botanical and Indian name.—Analysis and description of the grain.—Numerous varieties under cultivation in China, Japan, Burmah, Bengal, the Himalayas, Upper and Northern India, the Punjab, and Afghanistan, including Cashmere.—Mode of Cultivation.—The Afghan method of sowing.—Malted and distilled into spirits, called *Sham-shu* in China.—Similarly distilled in Nepaul, and consumed by the Nepaulese.—In Ceylon and Southern India, rice spirit is called *Arrack*.—Rice flour.—Its conversion into *Chuppatties* by Pogson's process.—Chinese method of preparing steamed rice bread or cakes.—Chinese method of manuring or enriching their blood by boiling their rice in water, holding the soluble salts of rice-husk ash in solution.—The food phosphates all present in such ash, though not present in husked rice.—Undoubted value of the system, as proved by the build, vigour (mental and bodily), and strength of the Chinese.—Value of rice husk as fuel.—The ashes so rich in pure transparent silica as to make glass.—The straw used as fodder.

212. *Can you explain what is meant by the word rice?*

Yes. When the husk is removed from the ripened grain of the rice plant (which in its natural or unhusked state is called "Paddy") the kernel which remains bears the name of rice and cleaned rice.

The botanical name of the rice plant is *Oryza*. Two species are known: 1st, the *Oryza latifolia*, or broad-leaved rice; and 2nd, the *Oryza sativa*, or cultivated rice. The Hindee name for unhusked rice is *Dhan*, and the husked rice is called *Chawul*, whilst boiled rice is called *Bhath*.

The native agriculturists have local names for each kind of rice cultivated by them: thus there is early rice, second early, and late rice. The rice cultivated in the Himalayas, up to an altitude of 3,000 feet above the sea level is called hill rice. The grain is cultivated in Cashmere and various parts of Afghanistan.

NOTE.—"The Chinese from time immemorial have been remarkable for the superior quality of their rice and rice culture. The Japanese also possess superior rice, one variety being so rich in vegetable fat that the Chinese will not eat it. This valuable variety should be introduced into India, and would be highly prized by all classes, who, when able, enrich their boiled rice with *Ghee*, the Hindee name for stale and more or less rancid boiled butter."—*J. F. Pogson.*

213. *Can you give a description of the grain, and of its components?*

Yes. "The seeds when husked are one to three lines long, white, nearly transparent, cylindrical, linear, furrowed, obtuse at both ends, brittle. Flour very white, friable, insipid, dry, inodorous and tasteless, very soft, and swelling much when cooked."—*O'Shaughnessy.*

According to Vogel's analysis, rice contains in 100 parts, starch 96, sugar 1, fatty oil 1·5, and albumen 0·2.

NOTE.—"According to Professors Johnstone and Church, the composition of rice is, water 14½, fibrine, 7½, starch 76, fat ½, fibre 1, ash 1—total 100."—"*Chemistry of Common Life*," p. 76.

214. *What soil, manure, and peculiar method of cultivation does the rice crop require?*

Clay soils which may easily be flooded are, as a rule, selected for rice cultivation.

The Indian zemindar does not manure his rice field; but the Chinese agriculturist does so, using human manure in solution. The Indian[1] zemindar raises rice in seed beds, and when sufficiently grown the seedlings are transplanted into the flooded rice field, which is constantly kept under water. The rice is ready to cut in Upper India in all October, and much earlier in Bengal and elsewhere, according to the nature of the climate and seasons.

The Afghans soak their seed rice in water until the grain sprouts. It is then mixed with *quantum suff.* of ashes to absorb superabundant moisture, and the seed rice so prepared is sown broadcast over the surface of the rice field. The seed is harrowed into the field without delay, and the field is flooded; germination speedily follows, and the young crop is irrigated as often as necessary until it is sufficiently grown, when the land is kept under water till the ears ripen and the grain is harvested.

NOTE.—"The Afghan system is deserving of adoption, as it does away with all chances of catching fever; whereas under the Indian plan almost all concerned in transplanting rice seedlings are, as a rule, smitten with fevers of more or less virulence. The rice harvest in the Himalayas is also attended with fevers, owing to the

[1] In some parts of Upper India the seed rice is sown broadcast on dry land. It is then harrowed in, and the field is flooded.—*J. F. P.*

soil being water-logged and giving off offensive gases, due to the death and decomposition of innumerable forms of aquatic insects and microscopic organisms. It is more than probable that the birth and death of successive generations of aquatic insects supplies the standing rice crop with suitable manure."—*J. F. Pogson.*

215. *Besides being used for food, can husked and unhusked rice be turned to any other remunerative account?*

Yes. Malted rice, when distilled, yields a very strong and pure spirit, called *Sham-shu* in China, and *Arrack* by Europeans in India, the word being a corruption of the Hindee word *Urruck*, meaning essence.

The beer made from malted rice is of the best description and highest quality; and as rice contains 20 per cent. more of starch than does the best wheat, the advantage of converting rice, when cheap and plentiful, into starch, for export to the United Kingdom, is apparent.

216. *Can you explain the methods by which rice flour is made into bread?*

Yes. The Chinese knead up rice flour with hot water and heat the dough sufficiently to form a plastic mass, which is rolled into thin cakes like Indian wheat-flour *Chuppatties.* A vessel of suitable size containing boiling water has a piece of stout calico tied over its mouth; a rice cake is placed on the calico until the under side is cooked by the steam, when the cake is turned and the other side similarly cooked, when the cake is removed and another put in its place; the process being repeated till all the rice dough is converted into steamed rice cakes or bread.

The second, or Pogson's process, produces a dough

which, when rolled into thin cakes and baked on an iron disc (called *Towa* in India, and made for the sole purpose of converting dough cakes into unleavened bread, called *Chuppatties*), gives a very palatable bread; which was highly appreciated by a Government elephant, who ate them as fast as offered. If the elephant had refused to eat the bread, it was to have been considered as condemned. There was a valid reason for adopting this curious test. Elephants are daily supplied with thick cakes made of wheat flour, and cooked on the *Chuppattie* plan; and as the elephant will not touch any food which his wondrous instinct tells him is to be avoided, the appreciation of the warm rice cakes by the elephant showed that it was safe and palatable. A horse was then offered the bread, and with like results. The inventor and members of the official committee did the same, and expressed approval after partaking thereof.

The rice bread is prepared as follows, viz., to the rice flour is added and mixed together in the dry state three per cent. of powdered and sifted bark of the *Laurus malabathrum*, vel *Cinnamonum malabathrum* tree. This bark is sold in all Indian bazaars, and its Hindee name is *Tuj*. It has the smell of cinnamon, and the powder when infused in water produces a rich mucilage, which possesses a very remarkable affinity for sugar. On water being added to the rice flour so enriched, a dough is at once obtained (owing to the affinity of the pulverized bark for starch), which, on being sufficiently kneaded, may then and there be converted into cakes or rice-flour *Chuppatties*. It is highly probable that if some sugar was to be added to such dough, and, after it was worked in, the proper quantity of yeast was incorporated therewith, that the dough would rise, and, when baked, yield a very white, spongy bread.

NOTE.—"Rice is so cheap and abundant, that if the proper percentage of gluten obtained from pea or bean meal was to be mixed with rice-flour dough, prepared as above, the resulting baked bread would be just as nourishing as wheaten bread. One hundred lbs. of pea or bean meal contain 24 lbs. of gluten, which may very easily be separated from the starch with which it is associated, and thus render the regular manufacture of rice-flour bread a new and profitable industry. Biscuits of a highly nourishing character could be made by mixing bean or pea meal (both peas and beans being previously parched on the Indian plan in hot sand) with rice-flour dough, made as indicated."—*J. F. Pogson.*

217. *The Chinese are a rice-eating nation; they are strong and sturdy: can you explain how they cook rice, and thrive on a perpetual rice diet?*

Yes. The Chinese set fire to rice husk, and the resulting ashes, which are very rich in pure transparent silica, also contain all the food phosphates, which, though absent from the kernels of rice, are by Nature's decree stored in the husk. This rice-husk ash, on having water poured over it, produces a solution of the food phosphates. In this aqueous solution the rice is cooked, just the proper quantity of solution being used to thoroughly cook the rice. Hence there is no rice water to be thrown away. The result of this admirable system is that everything present in the raw rice is present in the cooked rice, *plus* the food phosphates obtained by cooking the rice in prepared water which holds them in solution.

This method of daily manuring the blood fully accounts for the strength and virility of the Chinese nation.

NOTE.—"The population of Bengal proper are, like the Chinese, a rice-eating race; but they boil their rice in ordinary water, and eat the boiled rice, and drink the water in which it was boiled. They know nothing about the food phosphates present in rice-husk ashes, and there can be very little doubt that the stature and deficient strength and stamina of the Bengalee is due to the want of food phosphates."—*J. F. Pogson.*

218. *Has rice straw and rice husk any economic value?*

Yes. Rice straw is used as fodder and rice husk answers well as fuel. The ashes, being rich in silica, when fused with soda, yield a crude glass, with which the glass bangles worn by Indian women are made.

NOTE.—"The analysis of rice-husk ash, made for the writer in 1867, has been lost. A new variety of rice, which is cultivated on hill sides and does not require irrigation, has been brought to notice in the *Indian Agriculturist* for August, 1882. *Vide* page 281, called 'Dry Paddy.'"—*J. F. Pogson.*

CHAPTER IX.

MAIZE.

Botanical and Hindee name.—Its extreme antiquity in India.—Analysis and description of the grain.—Its antiquity in America so very remote as to be untraceable.—The numerous varieties of maize in America, quite unknown in India, till introduced by Europeans, and the Agri-Horticultural Society of India.—Varieties of maize under cultivation in America and in India from imported seed.—Mode of cultivation as practised by Indian Zemindars.—Improved, and proper mode of cultivation, not yet adopted by the Zemindars.—Successful cultivation of maize in Kotgurh, and also in Kulu, by the Reverend Mr. Carleton, American missionary.—Analysis of the ashes of Maize.

219. *Can you explain what is meant by the word Maize?*

Yes. Maize is derived from the botanical name, *Mays*, or *Zea*; from *Zea* we have *Zeine*, of which presently. Its Hindee name is *Mukkee*, the cobs are called *Bhotta*. The ripe seeds, when removed from the spike, are also called *Mukkee*. When these are ground into meal the product is called *Mukkee ka atta*, which means flour or meal of *Mukkee*. The seeds, when parched in a hot sand-bath, burst open and assume curious forms of a milk-white colour, due to the change which has taken place in the starch and gluten. The parched grain is in Hindee called *La-wah* and *Kheel*.

NOTE.—The cob of Indian corn, called *Bhotta* in Hindee, is repeatedly alluded to in the "Maha-Bharut," a poem which is admitted to have an antiquity of over two thousand years B.C. India, however, only possesses one description of maize, and that of inferior quality, due no doubt to degeneration. But America possesses numerous varieties, from the magnificent gigantic *Cusco* maize to the dwarf Golden Popcorn.

220. *Can you give a description of the grain of maize, and of its components?*

Yes. The seeds are rounded, crustaceous at the surface, smooth, naked, shining, of gilded, white, or purplish shades, disposed in a dense cylindrical spike, arranged in longitudinal ranks, and as if inlaid on the central axis or receptacle, parenchyme white and farinaceous. One hundred parts or pounds of Indian corn meal contain, of water 15 lbs., of gluten 9 lbs., of fat 5 lbs., of starch 64 lbs., of fibre or bran 5 lbs., mineral matters 2 lbs.

According to Professor O'Shaughnessy, the peculiar substance *Zeine*, which exists in Indian corn meal, when separated therefrom, is found to be "soft, malleable, elastic, like gluten—solid when cold, colour golden yellow, taste and smell peculiar, transparent in thin leaves, inflames at a taper, sp. gr. 1·0347, insoluble in cold water, softened but not dissolved by hot water, slightly by ether, soluble in alcohol and the oils, and is converted into a butter-like matter by nitric acid."

221. *Can you give the ordinary names of the various varieties of maize or Indian corn generally cultivated in America?*

Yes. But before doing so it is necessary to explain

that all sweet varieties of Indian corn are, in America, grown for family use, and for sale in the green state. Whilst other varieties are for field cultivation and popping, or parching.

The best sweet varieties are—1. Potter's Excelsior Sweet; 2. Marblehead Early Sweet;* 3. Pratt's Early; 4. Early Minnesota Sweet; 5. Early Narragansett; 6. Crosby's New Early Sweet;* 7. Forty days;* 8. Mexican Sweet;* 9. Golden Sweet; 10. Egyptian Sweet;* 11. Stowell's Evergreen Sweet; 12. Marblehead Mammoth Sweet;* 13. Tom Thumb; 14. Moore's Early Concord Sweet, and 15. Sweet Fodder Corn.

The best varieties for field culture are—1. Queen of the Prairie Dent;* 2. Chester County Mammoth;* 3. Cuzco, or South American Giant Corn;* 4. Sanford;* 5. Bailey's Ensilage; 6. Blunt's, or Prolific Field;* 7. Longfellow's Field;* 8. Adam's Early;* 9. Improved Early Yellow Canada; 10. Hundred Day's Dent; 11. Silver-laced Pop, and 14. Egyptian Popcorn.

NOTE.—"All the above varieties may be obtained from Mr. James J. H. Gregory, Marblehead, Massachusetts, United States, America. Illustrations have been given of the cobs of several varieties of the seeds named. The zemindars of India have no knowledge of these valuable varieties of maize, one of which, called "Blunt's Prolific Corn," will yield from three to six ears to the stalk; and as over one hundred bushels of shelled corn to the acre have been obtained from this particular variety, its cultivation in the northern parts of the Punjab is strongly recommended. The Rev. Mr. Carleton, missionary of Kulu, has raised this variety with great success. At the writer's request the Council of the

Agricultural Society of India obtained a supply of seed from Mr. Gregory, and the results will appear in the Appendix.

"The bushel of shelled Indian corn weighs fifty-six pounds. Hence the acre, under proper cultivation, will yield 5,600 pounds, or sixty-eight maunds and ten seers. The varieties of seed maize, marked with an asterisk, have been introduced this season (1882) by the Agri-Horticultural Society of India, and the writer has distributed his share to friends willing to give them a proper trial at Simla, Dehra Doon (three places), Rajpore, at the foot of the Mussoorie Hills, the Kownr,[1] Jainarain Sing, and Mr. Sri. Lall, secretary to the Agri-Horticultural Society of Bijnor, a district of the North-western Provinces. The results of their reports will be found in the Appendix. Want of ground has prevented the writer from growing these valuable varieties of maize at Mussoorie."—*J. F. Pogson.*

222. *How does the zemindar cultivate his crops of maize or Indian corn?*

He ploughs his land after the first fall of rain, and sows maize seed broadcast as soon as sufficient rain has fallen. The young growing crop is repeatedly weeded, and superabundant plants are pulled up by the roots, and taken away for use as green fodder. The crop in due time comes into bearing, and if the soil is good, two cobs are often produced on one plant, otherwise one cob is the usual or ordinary yield per plant. The cobs are small in size, and the grain on the spike corresponds. During the season the green cobs are

[1] Reports that all have matured seed, which has been preserved for sowing in 1883.—*J. F. P.*

roasted and eaten by all classes. When the harvest is gathered, the cobs are sun dried and stored, and at convenience the grain is removed by hand from each spike. The stalks, after being dried, are made into ricks. The dried stalks, when chopped into small pieces with a *Phursa*, is given as fodder to the zemindar's plough cattle. If he owns any milch cows, or buffaloes, the dry chopped fodder is placed in large vessels of pottery. Water is poured over it, and some oil cake broken up and mixed therewith, and when ready is given to the cows. The mess so made is called *Saunee*, whilst the dry stalks of the maize are called *Kurbee*. The grain is consumed by the producers, who sell the surplus. The grain would not sell in the London market owing to its inferior quality.

223. *Can you state what the zemindar should do to improve his maize crops, so as to produce a grain which would at once meet with sale in the markets of England and Europe?*

Yes. He should obtain and sow the seeds of American maize suited to his locality, climate, and latitude. He should use bone manure, or fossil phosphate of lime, and in place of sowing his seed broadcast, and then thinning out the growing crop, he should sow the seed singly in drills, at intervals of 18 inches for large varieties, and at less for the smaller kinds. The zemindar cannot do better than follow the American plan of maize cultivation.

224. *Can you state how maize crops are cultivated by American farmers?*

Yes. The American farmer selects a warm and rich soil for his maize crops. After it has been ploughed

EGYPTIAN SWEET CORN.

MAIZE.

1536

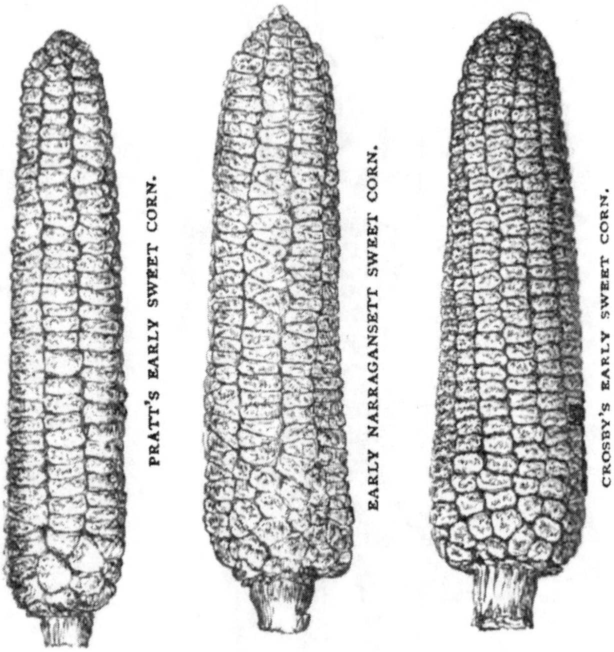
PRATT'S EARLY SWEET CORN.
EARLY NARRAGANSETT SWEET CORN.
CROSBY'S EARLY SWEET CORN.

MAIZE.

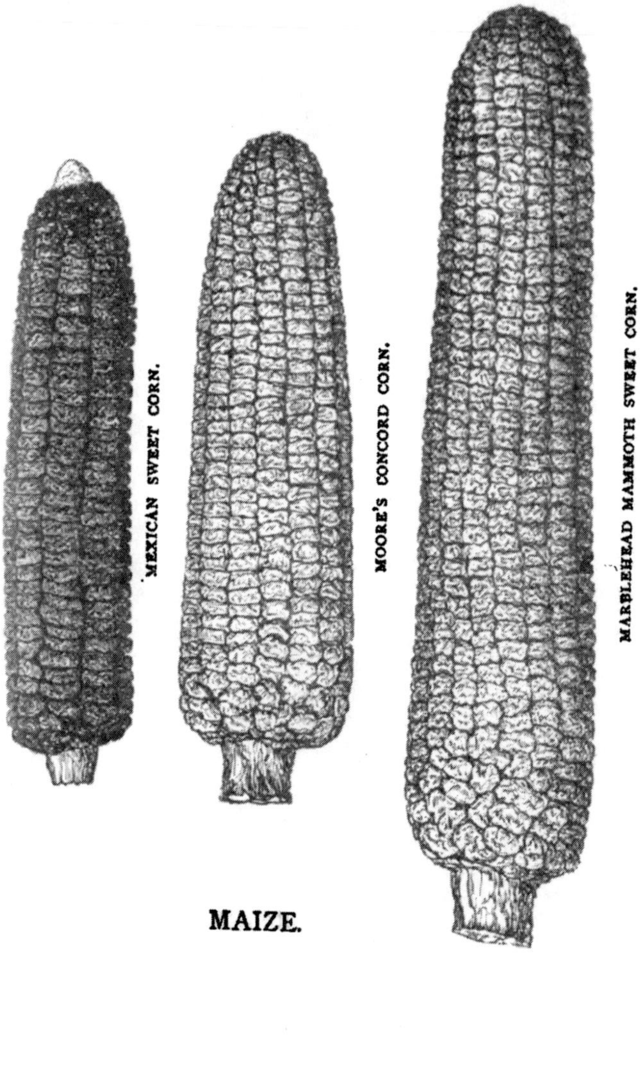
MEXICAN SWEET CORN.
MOORE'S CONCORD CORN.
MARBLEHEAD MAMMOTH SWEET CORN.

MAIZE.

153*d*

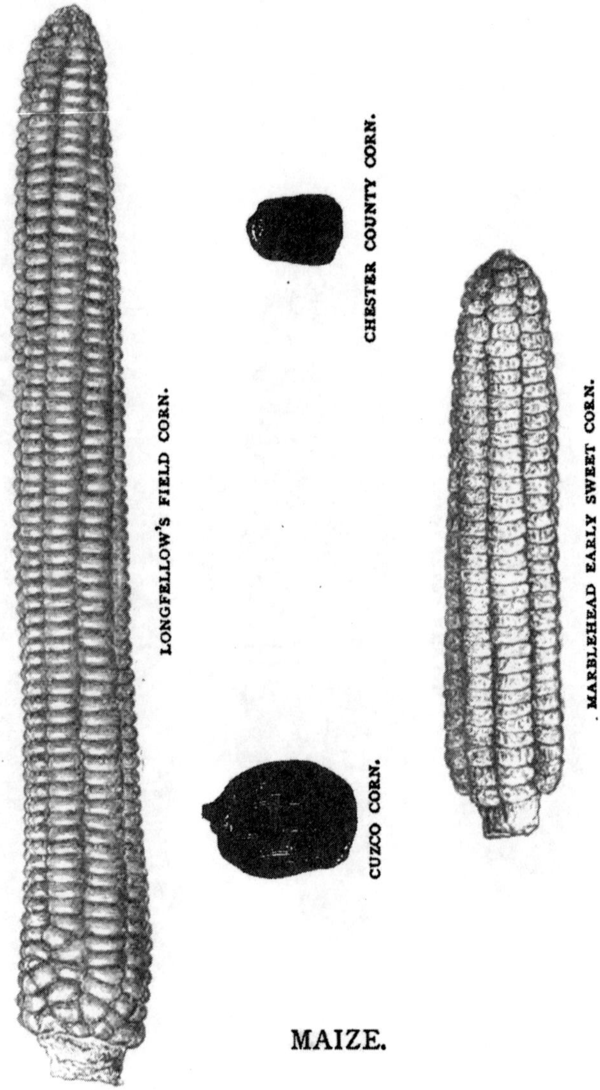
LONGFELLOW'S FIELD CORN.
CHESTER COUNTY CORN.
CUZCO CORN.
MARBLEHEAD EARLY SWEET CORN.

MAIZE.

and pulverized, he adopts drill cultivation, as it is more profitable than mound cultivation. For the larger varieties of maize, the drills should be 2½ feet apart, and the stalks, or plants, 1 foot apart in the row, and for the largest varieties the drills should be 3 to 4 feet apart, and the plants 18 inches apart in the rows. Rich manure is to be used in the drills, and not scattered over the surface of the entire field previous to sowing. The field is to be weeded as often as necessary.

NOTE.—"The writer has very successfully grown numerous varieties of American maize, from the magnificent Cuzco down to the Dwarf Popcorn. The seeds of all the very large varieties were sown singly at intervals of 18 inches, the Cuzco excepted, to which 24 inches was allowed. The manure used was made from bones and phosphatized limestone and other substances. Half a teacup full of this dry manure was dropped at 18 inches interval, and dug into the soil, and a single seed sown in each circle so manured. The plants grew vigorously, many attaining the height of 14 feet. Three to four cobs were borne on most of the plants, whilst all had two cobs. The smaller varieties throve as well. The culture was carried on at Simla, and during 1879 and 1880, at Kotgurh. The seeds of all kinds (Cuzco excepted) were distributed to zemindars at both stations, and the superior varieties of American maize cobs which have for years been sold in the Simla market are the produce of acclimatized seed, distributed to zemindars during the years 1860 to 1865. The Kotgurh zemindars carefully preserved all seed obtained from the crop of 1879 for future sowing. In 1880 the number of new applicants for American maize seed was so great that the demand could not be met. The facts stated will show that the hill zemindars fully appreciate the value of superior maize seed. In the plains of India, the early kinds only will suit certain latitudes, as the zemindar wants his land to be free for wheat sowing in September to October. West of the river Jumna, and on to the Indus river, the later varieties would best suit the soil and climate, as the wheat crop is not sown till November. Blunt's Prolific Field Corn would ripen perfectly west of Lahore, whilst east of the Jumna the same maize would be unripe in September to October, hence unfitted for the latitude. In the Simla hills, including Kotgurh, a maize that will not ripen seed at an altitude of 6,000 feet above the level of the sea, will do so perfectly at 5,000, or lower, as the case may be. Hence the great value of the locality for raising acclimatized seed from imported stock, for subsequent distribution in the Plains."—*J. F. Pogson.*

225. *Can you explain what is meant by the words "rich manure," as used by the American farmers?*

Yes. It means a manure which, in addition to various kinds of animal dung, contains a considerable quantity of all the mineral matters present in Indian corn ashes, and also in the leaves and stalks, and of these potash, magnesia, and phosphoric acid are the chief. Lime is needed for the entire plant, as the development of a rich sap in the thick stalks cannot take place without the aid of lime. Therefore, if the soil be poor in lime, it must be added, and the best and most economical way to use it is to allow one dessert spoonful of slaked lime (Bazar, *Chuman*), to each manured circle, before sowing the seed.

NOTE.—"The Kotgurh soil being very poor in lime, it was added to the mineral manure, which answered all purposes, without the aid of cow and horse dung."—*J. F. Pogson.*

226. *The river Sutledge divides the valley of the Sutledge into two parts; on the right is the grand slope on which the station of Kotgurh is built, on the opposite slope lies Kulu and the Rev. Mr. Carleton's mission station. Has maize from American seed answered expectation in his locality?*

Yes. So much so that the Punjab Government presented that gentleman with 500 rupees as a reward for his very successful cultivation of maize, Blunt's Prolific Field Corn being one of the varieties.

Composition of the Ashes of the Grain of Indian Corn.

Potash and Soda	32½
Lime	1½
Magnesia	16
Oxide of Iron	0¼
Phosphoric Acid	45
Sulphuric Acid	3
Chlorine	0¼
Silica	1½
	100

REMARKS.

No analysis is available of the stalks and leaves of Indian corn.—*J. F. P.*

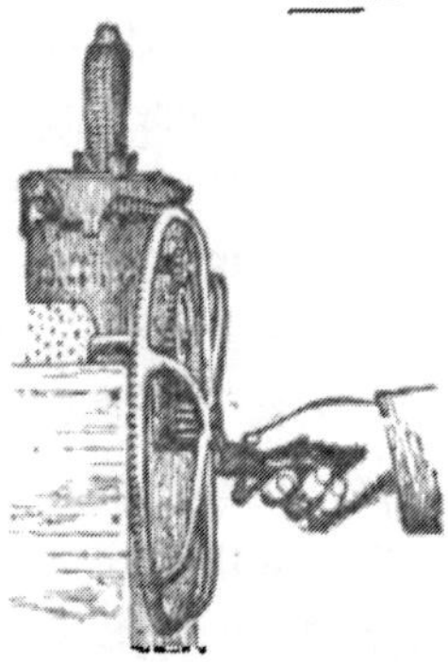

PEERLESS CORN SHELLER.

CHAPTER X.

SUGAR-PRODUCING SORGHUMS.

Introductory observations on the Chinese sugar-cane, or *Sorghum Saccharatus Sinensis;* African sugar *Sorghums;* the American Sugar Sorghums, called the Early Orange Cane; and the Early Amber Sugar-cane.—These varieties described.—Soil and manure needed for their proper cultivation.—Sugar, how made from the expressed sweet juice.—The great value of these two *Sorghums* to the Indian zemindars of the Himalayas and Plains of India explained.

INTRODUCTORY OBSERVATIONS.

THE Chinese have from very remote times manufactured sugar, and the far-famed Chinese sugar candy, from sugar-producing *Sorghums.* In India we have the red, the black, and the bi-coloured varieties of Sorghum. This last being called *Impee* indicates its Chinese origin; the other two are from some part of Africa.

The Early Amber Sugar-cane (*Sorghum*) has this year (1882) been introduced into India by the Agricultural Department, as also by the writer. The Early Orange Cane (*Sorghum*) has yet to be introduced.

The three sugar Sorghums, when first introduced into India many years ago, were looked upon as plants capable of producing an abundant supply of superior green fodder for Government elephants and horned cattle, and were cultivated accordingly, being sown in spring if irrigation was available, otherwise when the rains commenced, or at the same time that the ordinary non-sugar-producing *Sorghum* (the *Holcus Sorghum vulgare* of botanists) of the country was sown. The consequence of this plan of procedure was that the real value of the sugar Sorghums remained unknown for several years. In the spring of 1879 the writer obtained a small packet of seed of *Sorghum bicolor* (Chinese name, *Impee*,) and of it a portion was sent to the Rev. Mr. Carleton, in Kulu, and a portion of the remainder was sown in the writer's garden, as one of several agricultural experiments, having for their object the improvement of the agriculture of the Simla sub-district of Kotgurh.

The seeds sent to Mr. Carleton germinated freely, and the half-ripe ears of Sorghum sent over for inspection were far superior in size to those growing at Kotgurh. The unripe though fully formed grain in the ears was much larger than those on the Kotgurh plants, occupying a square rod of ground. Mr. Carleton had no idea that he had been growing a valuable sugar-producing Sorghum, and the writer was no better informed.

When the seeds were nearly ripe, the remarkable likeness of the stalks to sugar-cane with very long joints induced the writer to cut one of the thickest; the peculiar smell of sugar-cane when cut was apparent. The cane was peeled and cut up like sugar-cane for chewing. The first mouthful proved that the juice was as sweet as that present in the best sugar-cane of the Plains of India.

The discovery was an important one, and showed conclusively that a cool Hill (Himalayan) climate best suited the *Impee*, if the object in view was to cultivate it as a sugar-producing Sorghum. To ascertain how the standing crop would bear the cold of the locality, it was allowed to stand the best part of November, 1879, and only cut down when a snowfall was expected.

A dozen joints of the *Impee* (*Sorghum*), measuring ten to twelve inches each, were sent by post to Calcutta, for submission to the Council of the Agri-Horticultural Society of India. The rest of the crop (after the seed ears had been gathered) afforded a very considerable treat to the immediate population, irrespective of caste, creed, and colour, old and young included.

In his Report to the Agri-Horticultural Society, the writer pointed out that the *Impee*, being a sugar Sorghum, the proper time for sowing its seed in the Plains was in October, as a cold temperature was required to develope an abundant supply of rich saccharine juice, as proved by the Kotgurh experiment, where the nights in April, May, June, and July are cooler than the nights in the Plains of India during the first half of October, whilst the Hill temperature of October and November was equal to that of December in the Plains of India. Hence it was established that if the *Impee* bore the cold of the latter months in the Hills, it would withstand the much milder cold in the Plains, and therefore it was a decided mistake to grow the *Impee* as a hot-weather and rains crop for fodder, when it should have been sown and grown during the cold season for the express purpose of affording canes rich in saccharine juice, and only requiring pressing and boiling, to yield superior sugar and molasses.

The remarkable success with which the American

farmers have grown the two varieties of sugar-producing Sorghums in a climate by no means resembling that of India during the hot season and rains, proves conclusively that, if India is to become a Sorghum sugar producing country, the time and season of sowing must be changed from the hot to the cold season in the Plains. It stands to reason that the various districts of Upper India (the Punjab included), which yield the best crops of sugar-cane in the second year of its growth, will also under proper culture yield the best crops of sugar-producing Sorghums within eight months of seed sowing; under these conditions the sugar Sorghums will give two sugar harvests to one obtained from the sugar-cane. This should satisfy all concerned of the very profitable nature of sugar Sorghum cultivation when properly conducted.

227. *Can you state what description of soil and manure is needed for the proper cultivation of sugar-producing Sorghums?*

Yes. The American farmer selects a rich calcareous soil for sugar Sorghum cultivation, and he uses manures rich in phosphates, as also lime and bone dust. The Indian zemindar should select the best sugar-cane soil for sugar Sorghum cultivation, and use similar manures with potash and sulphate of iron in addition. Sugar-cane manure will answer as well for sugar Sorghums.

228. *How should the zemindar prepare the land, and how sow the seed of the Impee, the Early Amber Sugar-cane, and the Early Orange Sugar-cane?*

He should plough and prepare the land as if for sugar-cane. Then mark it off into beds three feet square, and in the centre of each bed make a circle twelve inches in diameter. This could easily be done by upsetting a twelve-inch basket on the spot and with a peg marking out the circumference. As this was done, a second man should drop as much sugar-cane manure as would fill a teacup into the marked circle, and a third man should dig it into the soil. By this plan 4,840 circles, three feet apart every way, from centre to centre, would be laid out, and be duly prepared for sowing.

The circles being ready, ten seeds should be sown in each in quincunx order, two seeds in each dibble. The square within the circle should measure nine inches on each side. Two seeds should be sown at each of the angles, and two seeds in the centre of the square—or ten seeds in all. Thus the acre would require 48,400 seeds. All seeds that fail to germinate should be replaced by fresh sowings, so that 48,400 plants per acre should be present as seedlings. After the plants are nine inches in height, soil should gradually be piled about them, so as to make a mound, on the flat surface of which the ten plants would be growing. The object in view is to prevent the plant roots from being submerged by an unexpected rainfall or too much canal irrigation. The best way to guard against this is to make furrows in the line of ascertained drainage, and to pile the soil taken out of the furrows about the plants. Cross furrows at right angles to the first set should then be made, and the soil similarly disposed of, with the result of having an open-drain furrow between each set of plants, so that, happen what may, perfect drainage would be secured. If the land and crop required artificial irrigation, water could be let into the furrows till all were half filled, when the closed exits should be opened, and all superabundant water drained off. In this manner the soil would be kept moist, without the land becoming water-logged or the plants being inundated.

All weeds should be removed, and the field kept clear of them.

Most of the plants will throw up two stems or stalks, others more. All should be allowed to grow.

When the seed in the ears is almost ripe, the canes are fully ripe for cutting and pressing.

229. *What should the zemindar do to obtain the largest quantity of saccharine juice from his Sorghum canes?*

He should cut off the seed ears as soon as they are fully developed, and before any seed has formed in them. The result will be that the canes will be very much richer in saccharine juice, and will mature quicker than if the seeds were allowed to ripen, and so convert a considerable quantity of sweet juice into starch. If the zemindar declines to sacrifice the seed crop, then he must rest content with a smaller yield of sugar and molasses.

230. *Can you state what quantity of sugar the American farmer obtains from each gallon of syrup, and how many gallons of syrup from an acre of land under sugar Sorghums?*

Yes. In the state of Minnesota the yield of sugar is from five to six pounds from one gallon of syrup, which weighs thirteen pounds and four ounces, and the yield varies from one hundred and twenty-five to one hundred and fifty gallons of syrup per acre.

On this subject Mr. James J. H. Gregory, of Massachusetts, states: "A great success has been reached at last in the manufacture of syrup up to latitude 44° 30′ N, which includes almost every portion of the Northern States. The yield per acre of syrup is from one hundred and forty to two hundred and eighty gallons, and the produce of sugar is about six pounds to the gallon of syrup. This syrup is equal in quality to the best syrup of the stores, and brings as high a price in the market."

NOTE.—In reference to the manufacture of sugar from the saccharine juice of the two American sugar Sorghums, the reader is referred to the standard work of Mr. J. A. Hedges, which gives full instructions for the manufacture of sugar. This work may be obtained through Messrs. Thacker, Spink, & Co., of Calcutta, or from Mr. James J. H. Gregory, Marblehead, Mass. Price $1.90.

231. *What should the Indian zemindar do in the matter of sugar Sorghum cultivation, until seeds of the two American varieties become generally available?*

He should sow the seeds of the *Impee* on the same plan as indicated for the American Sorghums, and convert the expressed sweet juice into *Goor*. The red and black varieties should also be sown at the same time, and similarly manured and cultivated. If on trial their sap or juice was found to equal that of the *Impee*, more extended cultivation would follow ; but if not, the *Impee* would be the zemindars stand-by, and would afford him a profitable income from crude sugar, or *Goor*, and also a fair income from the sale of the canes, for eating like sugar-cane. Separate plots of land should be sown with sugar Sorghums, and the ears allowed to ripen seed for future sowing.

CHAPTER XI.

COMMON, OR NON-SUGAR-PRODUCING, SORGHUMS.

Indian name.—The varieties cultivated in India.—Identical with the *Doora* of Northern Africa.—The meal of the grain as nutritious as wheat.—Cultivated all over Hindoostan during the hot season and rains.—American, or Balding's Branching Corn, also called *Doora*, proved to be a hitherto unknown variety of *Sorghum*.—Introduced into India this year (1882) by the Agri-Horticultural Society of India.—The plant described.—Owing to its great productiveness, will be invaluable to the ryots and zemindars of India.

232. ***Can you explain what is meant by the botanical name, "Holcus Sorghum vulgare"?***

Yes. It is the scientific name of a cereal very largely cultivated all over India. The Hindee name for it is *Jowar*. The Bengalee name is *Dao Dhan*, and it is known by other local names in the Madras and Bombay Presidencies.

The grain is thus described by Professor O'Shaughnessy: "*Holcus Sorghum vulgare*, a native of India; seeds very hard and rounded, very variable as to size, sometimes entirely smooth; flavour insipid, parenchyme farinaceous; but little differing from maize in chemical composition. The stalks are sweetish."

NOTE.—"When the crop of grain is gathered, the stalks, with leaves attached, are cut down, dried, and stacked, and used as dry fodder as required. It is also called *Kirbee.*"—*J. F. Pogson.*

233. *How many varieties of H. Sorghum are cultivated by the zemindars of India?*

Three; the white, the red, and the yellow. The women grind the seed into meal, which, after being sifted, is kneaded into dough and made into unleavened cakes, which are most wholesome and nourishing, and constitute the principal food of the agricultural classes. The flour of the white variety is most esteemed. The size and productiveness of the grain depends on the soil, manure, and method of cultivation.

NOTE.—"This grain (*Jowar*) is the staff of life over nearly all Northern Africa, and is called *Doora,* which is its Arabic name, and means the pearl, or like the pearl. Of it (*Doora*) Professor Johnstone states: 'Though so small a seed, it is so prolific that it yields from 60 to 100 bushels an acre;' and under proper cultivation it should do so in Hindoostan."—*J. F. Pogson.*

234. *Can you state if any other variety of H. Sorghum is known to exist out of India?*

Yes. In the United States of America the farmers grow a variety quite unknown until this year (1882) in India. The grain or seed, strange to say, is also called *Doora,* as well as Broom Corn; whilst in Massachusetts it bears the name of Balding's Branching Corn.

235. *Can you describe this particular variety?*

Yes. The seeds exactly resemble in form the *Jowar* seed of India. But there is a very remarkable difference between them. The Indian varieties are of uniform colour, whilst the American *Doora* has a circular black patch, or eye, at one end of the seed, the rest of the seed being of a dull white colour. In America this variety of *Sorghum* is stated to yield from 18,000 lbs. to 30,000 lbs. weight of grain per acre. Of it Mr. James J. H. Gregory, of Massachusetts, states: "I find this grows half a dozen stalks from a single seed, yielding from 6 to 18 medium-sized ears. The stalk in height and bulk is enormous."

NOTE.—"This most prolific and valuable cereal has been introduced into India this year (1882) by the Agri-Horticultural Society of India. The writer's share of the seed has been distributed to five persons, two in the Plains, two in the Doon, or Dehra, and one at Rajpore, at the foot of the Mussoorie hills. The results will appear in the Appendix."—*J. F. Pogson.*

236. *How should all these varieties of Sorghum be cultivated?*

The instructions given for American maize are equally applicable to *Sorghum*. The American *Doora* will need ample manuring, and the manure to be used should consist of bone-dust, or fossil phosphate of lime, from the Sewallics, and *Kunkur* dust. The soil should be made rich in lime as well as in lime phosphate; wood and weed ashes should be liberally used.

237. *Can you state why the ryot and zemindars of India should take the American Doora, or Balding's Branching Corn, into especial and careful cultivation?*

Yes. The zemindars and ryots need a cereal which is at once exceedingly prolific and good as a bread-stuff. The American *Doora* entirely answers these conditions and as one acre of it under proper cultivation will supply twenty adults with 2 lbs. of grain daily for twelve months, any outlay on suitable manures would be made good by the enormous yield of grain.

NOTE.—"The sole obstacle to the introduction of this valuable cereal on the large scale is the want of seed farms. That at Caunpore does not possess a suitable climate, and where that exists, in the upper and northern districts of the Punjab, there are no seed farms."—*J. F. Pogson.*

The extract given beneath is taken from the *Tasmanian Mail* newspaper, of the 3rd December, 1881, page 27.

"Says the *True Republican*, an Illinois (U.S.A.) paper: 'Broom-corn, a semi-tropical, is likely at no distant day to revolutionize the breadstuff of the world. A process has been discovered by which the finest and most delicious flour can be made from the seed to the extent of one-half its weight, and leave the other half valuable food for making beef and milk. The average yield per acre is 300 bushels, and in many instances 500 bushels, or 30,000 lbs., have been secured. Nor does it exhaust the soil, as Indian corn, from the fact that it feeds from the deeper soil and assimilates its food from a cruder state. It belongs to the same genus as the sweet cane, commonly known as the *Sorghum*, which as an article of food is growing rapidly in public esteem, and from the seed of which a most nutritious flour can be obtained.'"

CHAPTER XII.

SUGAR-CANE CROPS.

Introductory observations connected with sugar-cane.—Botanical name and Indian names of Indian varieties.—Non-Indian varieties.—Analysis of sugar-cane juice, and analysis of mineral matters extant in sugar-cane juice.—Soils best suited to sugar-cane cultivation.—Sugar-cane manures.—Extract from *Tasmanian Herald*, showing the value of the sugar and molasses obtained in the United States of America from the sugar-producing *Sorghums*.

INTRODUCTORY OBSERVATIONS.

THE Land Revenue of British India for the year 1879-1880 amounted to between twenty-two and twenty-three millions sterling, and during the year 1879 the quantity of sugar purchased and paid for by England amounted to the large sum of £22,344,555, the sugar costing rather more than £1 per cwt.

India is capable of supplying Great Britain with unlimited quantities of raw sugar, manufactured or crystallized sugars of three qualities, loaf sugar, cake crystallized sugar, and sugar candy. But in order to do so, British capital, enterprise, and trained skill are essentially necessary, and until these are introduced into India this vast and most remunerative industry cannot be carried out. It should be remembered first, that in no other part of the world is trained agricultural labour so cheap and plentiful as in India. Second, that the population of every village contains a number of experienced agriculturists who thoroughly understand sugar-cane cultivation, but cannot enter on it from want of the necessary funds. Third, sugar-cane cultivation cannot be carried out without artificial irrigation, and the magnificent Indian canals are fully available for this purpose. Fourth, sugar-cane requires certain mineral manures, all of which, though not in use, are present in vast quantities in different parts of the country, and close to open lines of railway.

When these important matters are taken into consideration, British capitalists will not fail to arrive at the conclusion that the sugar trade of the world would pass into the hands of the agricultural classes of India if they were supplied with the money needed to cultivate sugar-cane, which their European employer would purchase at a remunerative price and convert into sugar at his sugar mills and factory.

The past and present yield of saccharine matters obtained from the juice of the sugar-cane have now to be considered.

The second edition of Dr. Ure's "Dictionary of Arts, Manufactures, and Mines" was published in 1840, and under the head "Sugar" we are informed as follows, viz.: "One acre of good land in the southern districts of Bengal (proper) is reckoned to produce 56,673 lbs. weight of sugar-cane, which yields 3,477 lbs. of *Sheerah* (*Anglicè* syrup), or pot extract;" or in Indian weights, 691 maunds and 5 seers of sugar-cane, and 42 maunds and 16 seers of *Sheerah*, or syrup. This pot extract, when boiled down to the solid state of cake extract, gave 2,856 lbs., or 34 maunds and 32 seers of *Goor*, or cake extract.

Had the *Sheerah*, in place of being boiled down to a solid, been treated in a different manner by the professional Hindu sugar-makers, the resulting product would have been *Sukkur*, or granulated pale brown sugar. This, when refined, would yield *Cheenee*, or the more or less white powdery sugar of India.

This *Cheenee*, under suitable manipulation, yields crystallized cake sugar, very like loaf sugar, but in the form of a cake or disc about 2 inches thick and 18 inches or so in diameter. This is called *Thallee-Kee-Missree* in Hindee.

When this cake sugar is dissolved in a certain quantity of water and placed in properly prepared hemispherical vessels of common red pottery, it forms crystals on the cotton threads set within the vessels, and the bottom is covered with a deposit of crystals of sugar. The vessel is broken, and the sugar candy so manufactured is called *Koojae-Kee-Missree*, under which name it is sold in all the bazars of Hindoostan.

The information given above shows that from time immemorial the Hindu agriculturist has grown sugar-cane, which the professional Hindu sugar-maker has in due course made into raw sugar, white sugar in powder just like fine flour, cake sugar, and sugar candy; and as he makes his sugars without the aid and use of bone charcoal (an abomination in Hindu eyes), he possesses *a secret of manufacture* unknown to the European sugar planter. The superior crystallized sugars made in the Rosa sugar factory in the district of Shahjeehanpore, North-Western Provinces, and the sugars of Cossipore, which are of still higher quality, are carefully avoided by the Hindoos, owing to bone charcoal being employed in their manufacture. Hence it follows that if the European sugar-maker desires to sell his sugars to the Hindoos he must adopt their mode of manufacture; but if he does not care for local sales, then all his produce will have to be exported.

The Government of India has recently published statistics of sugar production in India, in which we are officially informed that in the North-Western Provinces the estimated yield in the several districts is variously returned between 15 cwt. and 22 cwt. of sugar per acre. By the term sugar is meant the cake extract, or *Goor*, already alluded to.

We are further officially told that, "Imperfect as the information is, it is sufficient to establish the broad fact that an acre of sugar land (*sic*) in India yields on an average considerably more than the 8 cwt. of unrefined sugar, at which the *Produce Markets Review* estimates the out-turn."

As no *Goor*, or cake extract, is ever exported to London for sale, the "unrefined sugar" above alluded to must mean the *Sukkur*, or raw sugar, of the Indian sugar market, and if such be the case, further comment is unnecessary. But if *Goor*, or cake extract, is meant, then if in place of more than 8 cwt. we allow 9 cwt. per acre, the falling off in the produce of saccharine matter per acre equals 16½ cwt., and if this is true, it must be admitted that the soil from improper culture has degenerated so immensely as to yield less than 9 cwt. of cake extract, or *Goor*, per acre, in place of 25½ cwt. of forty years ago.

In the West Indies, *Goor* is never manufactured. But the West Indian sugar planter obtains in a good season from 40 to 50 cwt., 2 tons to 2½ tons, or 54 maunds to 67½ maunds of granulated raw sugar, corresponding to the *Sukkur* of India. Yet in the official paper under notice we are told that "Messrs. Burrows, Thomson, Mylne, and Co. produce from 25 cwt. to 30 cwt. of crude sugar as an average yield per acre. This corresponds closely with the average of 22 cwt. given for many districts in the North-Western Provinces and Oudh, and the Government of India is inclined to think that it is as close an estimate as can be made, though considerably less than the 2 tons, or 2½ tons. which a sugar plantation in the West Indies is said to yield."

How the compiler of the official paper quoted should have fallen into the errors he has done, it is impossible to say. But the broad fact is patent that, whilst the West Indian planter grows sugar-cane, and produces raw or granular sugar, and molasses in large quantity per acre (2 tons to 2½ tons), the Indian zemindar at the outside produces 22 cwt. of cake extract, or *Goor*, which by a most extraordinary confusion of ideas, is made to take the place of granulated sugar corresponding to that of the West Indies; whereas the one is totally different from the other in every respect, *Goor* not being convertible into any description of sugar, whilst the granulated sugar of the West Indies and the *Sukkur* of India, on being refined, yields all the varieties of sugar enumerated by the writer.—*J. F. Pogson.*

238. *Can you state what is the botanical name of the sugar-cane, and give a description of the plant?*

Yes. Its botanical name is *Saccharum officinarum.*

Of it Professor O'Shaughnessy states: "Culms cylindrical, marked in spaces by knots, or large and swollen joints; odorous; medulla abundant, juicy, and saccharine."

239. *Can you give the names of the Indian varieties of sugar-cane, and by which they are known to the people of Hindoostan?*

Yes. In Bengal proper three varieties of sugar-cane are cultivated. The first is called *Cadjoolee.* It is a purple-coloured cane. It is esteemed very good for making sugar, but is unknown in the West Indies. The second is called *Pooree.* It is a light-coloured cane, yellow inclining to white, deeper yellow when ripe and grown on rich soil. This variety is much eaten raw. The third variety is called *Cullorah.* It thrives on swampy lands, is light-coloured, and grows to a great height. Its juice is more watery, and yields a weaker sugar also than the *Cadjoolee.* In the North-Western Provinces, Oudh, and the Punjab a thin variety of sugar-cane, called *Oo-kh* in Hindee, and *Ikh* in Punjabee, is almost exclusively cultivated for sugar manufacture. A thick white variety, called *Poundæ,* is largely cultivated for eating or chewing, after the hard skin or coat is peeled off. A third, reddish purple variety, called *Souta,* and *Lall Poundæ,* is also cultivated, on the small scale, chiefly for consumption in the raw state. The juice is said to be very sweet, but the cane is too hard for the Hindu sugar mill.

240. *Can you name the foreign, or non-Indian, varieties of sugar-cane?*

Yes. In the West Indies the first and longest known variety is called the *Creole,* or common sugar-cane, originally obtained from Madeira; the second variety of the plant is the Otaheitan; the third is the Batavian, or violet-coloured sugar-cane.

In Louisiana (United States) five different varieties of sugar-cane are cultivated. The one most esteemed is the striped cane of Louisiana. The Batavian cane and the violet varieties from Java are also highly esteemed.

NOTE.—"Many years ago the Otaheitan sugar-cane was introduced into Calcutta and its neighbourhood, and answered well. But it has not as yet reached the agricultural classes of Upper and Northern India, and never will until seed farms are established in suitable localities north of Meerut, when all the foreign varieties of sugar-cane named could be obtained and acclimatized, and in due course distributed to the zemindars, who would at once enter on their cultivation."—*J. F. Pogson.*

241. *Can you state what quantity of sugar may be obtained from* 100 *lbs. of sugar-cane properly cultivated?*

Yes. In Java, in Cuba, and in the West Indies the planters are now extracting and sending to market 10 to 12 per cent. of raw sugar from 100 lbs. of canes. The juice of sugar-cane varies in richness and quantity. But it may be reckoned as a fair average product when one pound of sugar is obtained from one gallon (English) of juice.

242. *Can you state what substances are present in sugar-cane juice?*

Yes. When the juice is obtained from fully ripe canes 100 parts of it contain on the average as follows, viz.: water 71; sugar 18; woody fibre and carbohydrates 9½; saline matter ½; gluten and nitrogenous matter ½; colouring matters ½.

243. *Can you state what mineral matters are present*

in sugar-cane juice obtained from ripe canes properly cultivated?

Yes. In ten gallons (English) of sugar-cane juice, specific gravity 8½ degrees *Baume*, there are 5¾ ounces of salt, which consists of

	Grammes.
Sulphate of Potash	17·840
Phosphate of Potash...	16·028
Chlorure of Potassium	8·355
Acetate of Potash	63·750
Acetate of Lime...	36·010
Gelatinous Silica	15·270
	157·253

Grammes 157·253 equal 5·57 ounces avoirdupois.

NOTE.—"The advocates of *Poudrette* as a sugar-cane manure, should ponder over the above analysis, which shows that 10 gallons of sugar-cane juice contain no less than 105.973 grammes of potash salts. If sugar-cane juice contains this very large proportion of potassic salts, what quantity thereof must there be present in the leaves and stalks of the plant?

"Can there be anything more intensely absurd than the sight of one set of highly-paid Indian officials bemoaning the decadence of Indian agricultural products, sugar included, whilst another equally well-paid set will not permit potassic manures to be used, because an all-wise Providence has associated nitrate of potash with common salt, which must not be touched lest the income from the salt monopoly may be lowered? All the potassic salts named in the analysis could be elaborated from the nitrate of potash by the roots of the sugar-cane plant.[1] But the Indian Commissioner of Excise says, 'Touch it if you dare, save on my terms,' and these, unfortunately, are prohibitive.

"Bone-dust and fossil phosphate of lime can alone supply the all-important phosphoric acid, without which the phosphate of potash cannot be produced. The Hindoo will not touch the first, and though he would willingly use the second, it is simply unattainable."—*J. F. Pogson.*

244. *Can you state what description of soil and manure is best suited to the sugar-cane plant?*

Yes. After many years of observation it has been fully established in the West Indies that "where sugar-canes grow on a calcareous marly soil, in a favourable season, the saccharine matter gets so thoroughly elaborated, and the glutinous mucilage so completely condensed, that a clear juice and a fine sugar may be obtained without the use of lime."

Such soils are very uncommon in India, hence it follows that the best manures to be used are those rich in lime and potash, with a sufficient quantity of iron and phosphoric acid. The plant should be manured in preference to the field.

NOTE.—"In order to enable the critical reader to draw his own comparisons and conclusions between sugar-cane cultivation and the cultivation of sugar-producing *Sorghums*, an extract is given beneath from the *Tasmanian Herald* of the 13th of May, 1882. It should

[1] In the Appendix will be found an official report, in which the great value of saltpetre, or nitrate of potash, as a mineral manure for *Sorghum Saccharatum*, or sugar-producing *Sorghum*, is demonstrated.—*J. F. Pogson.*

be remembered that *Sorghum* sugar can be placed in the market within twelve months, whilst in India the sugar-cane planted in the spring of 1882 will not be ready for cutting till September or October, 1883."—*J. F. Pogson.*

"Within a comparatively short period the *Sorghum* has come into very extensive cultivation in the Middle and Western States of America for the manufacture of sugar. It is also grown in the Southern States lying north of the sugar-cane zone. It grows to a height of from 12 feet to 14 feet in those regions, and yields nearly 3 tons per acre, and contains about 10 per cent. of sugar. In a recent issue of the *Rural New Yorker* it is stated that crystallized sugar can be obtained from *Sorghum* of as good quality as the ordinary brown sugar. This can be refined into a white sugar by the ordinary methods. It is also mentioned that one acre of this crop gave 710 lbs. of brown sugar, and 737 lbs. of molasses, in which there were 569 lbs. of uncrystallized sugar. This yield was worth over £18. An excellent rum can be made from the molasses. It may be added that, in order to obtain the most satisfactory results, it is necessary to cut the stems while yet green but fully ripe—that is, just when the seeds begin to harden and are nearly ready to fall. The plants at this time contain the largest percentage of crystallizable and non-crystallizable sugar, and the proportion between these has reached its most favourable stage."

EARLY AMBER SUGAR CANE.

CHAPTER XIII.

OIL SEED CROPS.

Introductory observations. — Linseed, Gingilie, white and black mustard; white, black, and yellow rape; and annual castor oil crops.

INTRODUCTORY OBSERVATIONS.

INDIA possesses numerous plants whose seeds, when pressed, yield oil fit for food, and burning in the open pottery lamp of Hindoostan, called *Chiraug* in Hindee. In the present chapter only those varieties of oil seed-producing plants which are annually cultivated by the ryot and zemindar are brought to notice, and they are detailed beneath, their botanical and Hindee names being given.

1. Linseed (*Linum usitatissimum*), Hindee name, *Ulsee* and *Teesee.*

2. Gingilie, White and Black (*Sesamum Orientale, alba et nigra*), Hindee name, *Suffaed Til* for the white, and *Kala Til* for the black.

3. Three kinds of Mustard: the White (*Sinapis alba*), Hindee name, *Suffaed Surson;* the Black, (*Sinapis nigra*), Hindee name, *Kala Surson.*

3. The Yellow (*Sinapis ramosa*), Hindee name, *Peela Surson.*

4. Rape (*Brassica campestris*), Hindee name, *Doo-an;* Urdu name, *Lahee;* and Punjabee name, *Tara-Meera.*

5. The Annual Castor Oil (*Ricinus communis*), Hindee name, *Irindee;* also *Raindee.*

NOTE.—"India possesses a perennial castor oil which grows to the size of a small tree. It is to be found in most gardens belonging to Europeans, but is not cultivated by the agricultural classes."—*J. F. Pogson.*

245. *What description of soil and manure is needed by all oil seed-producing plants?*

All oil-producing plants require a rich soil, in which lime and phosphate of lime should be fully present, as also iron and potash.

The proper manures needed by oil-seed crops are those which contain bone-dust, or bone-ash, lime phosphates derived from fossil bone and its matrix, magnesia, sulphate of iron, and nitrate of potash, with a little common salt. Nearly all these substances are present in various kinds of oil-cake, hence should be present in both soil and manure.

246. *Can you state what substances are present in ripe dry linseed?*

Yes. "Linseed in its dry state contains as follows, viz.: 11·265 of oil; 0·146 of wax; 2·488 of soft resin; 0·550 of a colouring resinous matter; 0·926 of a yellowish substance analogous to tannin; 6·154 of gum; 15·12 of vegetable mucilage; 1·48 of starch; 2·932 of gluten; 2·782 of albumine; 10·884 of saccharine extractive; 44·382 of envelopes, including some vegetable mucilage. It contains also free acetic acid, some acetate, sulphate and muriate of potash, phosphate and sulphate of lime, phosphate of magnesia, and some silica."—*Ure.*

NOTE.—"When the oil is removed from linseed by pressing, all the substances above named remain in the oil-cake, which is of itself a true food for cattle, whilst its jelly might be used as nourishing food for the sick and convalescent."—*J. F. Pogson.*

247. *Can you state what linseed oil is composed of?*

Yes. Of carbon 76·01 parts, of oxygen 12·64, and of hydrogen 11·35 in one hundred parts.

248. *Does the linseed plant produce anything else besides oil-producing seeds?*

Yes. It yields flax, by which term is meant the bast, or inner bark, of the *Linum usitatissimum*, which is spun into yarn for weaving linen in European countries.

In India the crop is only grown for its seed, which is largely exported to the United Kingdom and to Europe.

249. *Why does not the zemindar grow this crop for its flax or fibre, as well as for its seed?*

Because the plant to produce flax requires a temperate climate, not to be met with in the plains of Hindoostan. It would answer perfectly in the inner or superior ranges of the Himalayas, but want of cheap capital and cheaper transit renders this cultivation and the preparation of flax out of the question.

NOTE.—"Ireland would cease to produce superior crops of flax if its climate underwent a change and became like that of India from March to the end of September. Egypt has from the remotest antiquity been a flax-producing country, and the ancient Hebrews, especially of the tribe of Judah, spun flax and wove fine linen, and no doubt flax was grown largely within sight of Jerusalem. That linseed was introduced into Ireland at some very remote period of antiquity, from the East, cannot be gainsaid, and the honour of its introduction lies between the ancient Phœnicians and the Hebrews, who so far back as King Solomon's time traded in company with Britain for its tin. The fleets of both nations may have landed Hebrew colonists, who brought with them the onion, leek, and linseed, all Oriental plants."—*J. F. Pogson.*

250. *Can you describe the Sesamum (Hindee, Til) plant, and the peculiarity of the oil obtained from its seeds?*

Yes. "The stalks are straight, herbaceous, nearly cylindrical, hairy, about two feet high; leaves oval-oblong, with long petioles, entire toothed, the upper leaves alternate; flowers solitary, axillary, with short peduncles and linear bractes, capsules oblong, rather compressed, marked with deep furrows; seeds slightly oval, small, white or brown black, inodorous, having an almondy taste.

"A very sweet clear oil is obtained by expression from the seeds. This oil is much used as an article of diet, and also for frictions. For all purposes of medicine and pharmacy it is quite equal to the best olive oil. Sesamum oil is insoluble in alcohol, specific gravity 0·911. In India various palatable sweetmeats are made from the seeds, and also from its oil-cake."—*O'Shaughnessy.*

NOTE.—"Nine pounds of the seed will yield two quarts of perfectly sweet oil. The writer has grown this plant at Simla, and also at Kotgurh. Hence it follows that the *Sesamum Orientale* may be grown successfully in all parts of the United Kingdom where flax is cultivated. This crop is a very profitable one, and its introduction into England and Ireland would be most beneficial to all parties. The seeds should be sown in spring"—*J. F. Pogson.*

251. *How does the zemindar cultivate the three named varieties of mustard and the rape crop?*

After ploughing and pulverizing the soil, he sows the seed of each broadcast, in separate fields or blocks of land. The only manure the zemindar can use, if so inclined, is cow-dung manure. With proper mineral manures (*see* Linseed) heavy harvests of oil seeds would result, and an ample supply of oil-cake would be available for sale as cattle food. Oil-cake is too valuable to be used as a manure in India.

NOTE.—"The exportation of oil seeds annually deprives India of immense quantities of oil-cake, which could be retained, if oil mills were erected in favourable localities."—*J. F. Pogson.*

252. *Can you state what mustard seed oil is composed of?*

Yes. Mustard oil contains carbon, oxygen, hydrogen, and nitrogen, as also sulphur. In India this oil is extensively used by the natives for culinary purposes, and especially for cooking fish.

253. *Can you describe the castor oil plant, and the mode of its cultivation by the zemindars?*

Yes. Its fruits are, "glaucous, capsules partially rounded, thorny, of middling hardness, with three monospermous cells, indicated by three projections externally. The seeds are oval, obtuse at both ends, the size of a large French bean, compressed, shining, convex superiorly, more flattened below, marked with an evident ridge, rather angular, marbled with little brown points or incomplete and interrupted red lines. The seed is, moreover, carunculate, the kernel is covered with a sheath and a separable integument, the former being dry, brittle, insipid, of middling thickness. On removing these integuments we find a very oily and beautiful foliaceous kernel."—*O'Shaughnessy.*

The zemindar having prepared the land, sows single seeds in straight lines, 3½ to 4 feet between seed and seed, and the same distance between the lines. The seeds are sown in spring. The ground is manured about the plants. The growing crop has to be irrigated. The seed pods ripen during the cold months.

254. *Can you state what castor oil is composed of?*

Yes. One hundred parts of castor oil contain, of carbon 74·0 parts, of hydrogen 10·3 parts, and of oxygen 15·7 parts. Castor oil is entirely soluble in alcohol. This oil is much used for medicinal purposes, and in India as a lubricant for railway engines and locomotives.

CHAPTER XIV.

FIELD PEA CROPS, INCLUDING THE JAPAN PEA. BEAN CROPS.

Introductory observations regarding peas and beans.—All varieties of peas and beans thrive in the Himalayas.—Cattle beans unknown in the Plains.—Their introduction recommended.—Pea and bean soils.—Manure for pea and bean crops.—Mode of cultivation—Analysis of bean and pea ash.—Pea cheese of China.

INTRODUCTORY OBSERVATIONS.

THE zemindars of the Bengal Presidency cultivate an inferior black description of pea, which is chiefly given to sheep and pigeons. But except in gardens, for table use, the cultivation of peas as a field crop is almost unknown; yet all varieties of dwarf peas could be very successfully grown as a field crop, as also could climbing peas where brushwood and *Urhur Dall* as well as cotton stalks were available. In like manner all varieties of dwarf beans could be raised as a field crop, whilst the zemindars of the hills north of Dehra, if once supplied with bean seed, could produce all kinds in abundance, including all varieties of cattle beans.

The white haricot bean, large and small varieties, thrives most luxuriantly in the Himalayas; yet India is supplied with the small haricot dried bean by the United States of America!

255. *Can you state what description of soil is best suited to pea cultivation?*

Yes. Peas require a sandy loam, neither too moist nor too dry. In India sandy clay soils are very common, and when manured with *Chunam* or *Kunkur* dust (nodular limestone reduced to powder) they become fitted for pea culture.

256. *How should the land be prepared for sowing peas, and how much seed is needed to sow an acre?*

The land after being ploughed should be laid out in drills two feet apart. The lime manure should be applied to the drills and mixed with the soil. Bone-dust to the extent of five per cent. should be mixed with the *Chunam*, as also wood ashes or saltpetre.

The seeds of dwarf peas of all kinds should be sown in drills, two seeds at every six inches. From four to five bushels of seed per acre is a proper allowance. The Japan pea assumes the form of a bush from two to three feet in height. In the plains it should be sown after the rains cease. Drills should be made three feet apart, and a single seed should be sown at every three feet; the spot where the seed is sown being manured as for beans. The botanical name of the Japan pea is *Soya Hispida.* It is half pea and half bean in appearance, with singular leaves, and pods somewhat like the pods of the *Cajanus sativa* or *Urhur Dall* of the next chapter.

NOTE.—"Pea cultivation can be most profitably carried on in the vicinity of cities, large towns, and military cantonments. The cultivation of climbing peas is the same as for dwarfs, only the drills should be three feet apart to allow room for bushing. It is not advisable to sow dwarf beans broadcast, as both seed and manure are wasted."—*J. F. Pogson.*

257. *Can you state what description of soil and manure is best suited to bean cultivation?*

Yes. Beans require the same sort of soil as wheat, namely, heavy clays, and should be sown in drills three feet apart. Single seeds at six inches apart for common kinds and dwarfs. The scarlet runner seeds to be sown nine inches apart, and in the hills one foot should be allowed between seed and seed. The furrow should be six inches deep. The manure to be used should contain *Chunam*, with ten per cent. of its weight of bone-dust mixed with it. Beans require more wood ashes and saltpetre than do peas.

258. *When should peas and beans be sown, and what quantity of seed beans is needed to sow an acre of land?*

In the hills, peas sown in October are some six to nine inches in height before the snow falls. They are not affected by the cold, and make rapid growth in March, and bear in April. Fresh sowings take place in March, and so on to May. In the plains, peas are sown in October and November, and, where the spring is not very hot, also in December. It is the heat which affects peas, and not the cold. In the plains, beans are sown after the rains cease—between September and October. They are not affected by the mild Indian winter.

Four bushels of seed beans are sufficient to sow an acre of land.

NOTE.—"Climbing beans should not be sown in localities where brushwood is scarce. The vicinity of cities and towns is the most desirable situation for bean culture."—*J. F. Pogson.*

259. *Can you state what substances are present in peas and beans?*

Yes. One hundred lbs. of dry peas and beans contain of water 14 lbs., of husk or shell 10 lbs., of starch 45 lbs., of gluten, in the form of legumin, 24 lbs., of fat 2 lbs., and of mineral matter 3 lbs. Beans and peas are chiefly remarkable for the proportion of gluten they contain, and for their constipating quality.

260. *What is legumin?*

"Legumin is another albuminous compound or substance resembling white of eggs, and is found in large proportions in beans, lentils, peas, and other leguminous seeds, hence its name."—*Johnstone.*

NOTE.—"In India the chick pea, or gram, when parched, is considered to be more capable of supporting life, weight for weight, than any other kind of food. It is for this reason the Hindu traveller eats *Suttoo*, or parched gram meal, when on a journey."—*J. F. Pogson.*

260½. *Can you state what mineral matters are present in* 100 *lbs. weight of bean ashes?*

Yes. They contain of potash and soda 45 lbs., lime 8⅔ lbs., magnesia 6½ lbs., oxide of iron ⅓ lb., phosphoric acid 33 lbs., sulphuric acid 4½ lbs., chlorine 1¼ lbs., silica ¾ lb.

NOTE.—Pea and bean straw afford excellent dry fodder for sheep and cattle. In India the straw of gram is much esteemed for fattening sheep and oxen. The green leaves of the gram plant are used, when in season, for making a very palatable curry.

261. *Can you state how the Chinese manage to prepare cheese from peas?*

Yes. "The Chinese are in the habit of making a real cheese from peas. For this purpose the peas are boiled to a thin paste which is passed through a sieve and coagulated by the addition of solution of gypsum. The curd is treated like that formed in milk by means of rennet. The solid part is pressed out, and, with the addition of salt, is wrought into cheese in moulds. This cheese gradually acquires the smell and taste of milk cheese. It is sold in the streets of Canton under the name of *Taofoo*, and when fresh is a favourite article of food with the people."

NOTE.—The above is from the report of M. J. Itier, and, according to Liebig, affords "a remarkable proof of the true nature of vegetable caseine, the fact furnished being quite independent of chemical researches."

"From the context, *pea meal* is the substance operated on, the gypsum being dissolved in hydrochloric acid." —*J. F. Pogson.*

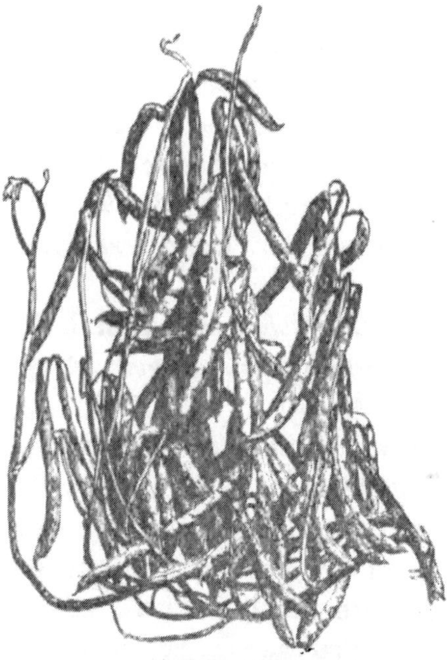

CHINESE BEAN.

CHAPTER XV.

DALL, OR PULSE CROPS.

The various pulses, called *Dall* in Urdu and Hindee, and largely cultivated all over India, for culinary purposes.—The charcoal of one variety used in the manufacture of gunpowder at the Government powder mills.

262. *Can you mention the Indian and botanical names of the various Pulses called* Dall, *and cultivated by the zemindars as an article of diet?*

Yes. The principal, and most popular *Dalls* are, first, the *Urhur*, or *Cajanus sativa;* second, the *Moong* or *Phaseolus radiatus;* third, the *Mussoor*, or *Ervum lens;* fourth, the *Koolthee*, or *Dolichos biflorus;* fifth, the *Ooord* (or a vetch); sixth, *the Mooth*[1] (or a vetch), *and* seventh, the *Chunna*, or *Cicer Arietinum*, commonly called Bengal Gram.

263. *Can you state when the* Urhur Dall *seed is sown, and why it is esteemed as a most nourishing description of food?*

Yes. It is sown in May or June, or when the rains commence, and the seed pods are ripe in March and April of the next year. The plant grows to a height of from six to eight feet, and bears an abundance of small pea-like pods, with round flattish seeds. These seeds

[1] Botanical name not known.—*J. F. P.*

when split, and cleared of their shell or husk, constitute the *Urhur Dall* of the bazar.

The shelled seeds when boiled in water afford a very nourishing gruel, or *Conjee*, for the sick, the weak, and convalescents. The porridge made with this *Dall*, either with or without curry spices, is a most palatable dish, and very sustaining and nourishing. It is for these reasons that this *Dall* is so highly esteemed by all classes.

264. *Can you state when the* Moong Dall *is sown, and why the better classes prefer it when cooked to other kinds of* Dall?

Yes. It is sown when the rains cease. This *Dall*, when shelled and split, has a pale yellow colour. It is very easily cooked, either with or without curry spices, and has a very delicate and palatable flavour. It is easily digested, does not produce flatulence, and is reckoned a cooling and nourishing food. It affords an excellent medium for the absorption and exhibition of *Ghee* (rancid butter, deprived of its water by boiling, and then preserved for use in pottery vessels), of which the rich natives are so fond; hence its popularity with the well-to-do classes. The cultivation is decidedly profitable.

265. *Can you state when the* Mussoor Dall *is sown, and why, as an article of diet, it is not so popular as the other two varieties?*

Yes. This *Dall* is also sown after the rains, and is harvested in the spring. When cooked, though very palatable, it possesses heating properties, which, however agreeable during the cold months, are the reverse during the hot weather months, or from March to the end of September. The *Revalenta Arabica*, so much in vogue some years ago, was made from the *Ervum lens*.

266. *Can you state when the* Koolthee *is sown, and the principal use made of this* Dall ?

Yes. It is sown after the rains. In the Madras Presidency the horses are fed on this grain after it has been boiled. In the Bengal Presidency it is never so used. The poorest classes cook and eat this *Dall.*

267. *Can you state when the* Oord *is sown, and how the grain is used?*

Yes. It is sown as soon as the rains commence, and it is harvested in October—November. This *Dall* is almost daily consumed by the vast agricultural population of the Bengal Presidency—throughout its length, from Behar to Peshawur. It is reckoned very nourishing and wholesome.

268. *Can you state when the* Mooth *is sown, and the principal use made of this* Dall ?

Yes. It is sown when the rains commence, and is harvested in September, October, and part of November, according to latitude. In the Bengal Presidency it is much used as cattle food, and is fattening. Horses out of condition, when fed on this *Dall* previously boiled, soon get fat. Sheep are also fed and fattened on the raw *Mooth.*

269. *Can you state when the* Gram, *or* Cicer Arietinum, *is sown, and the principal use made of this grain ?*

Yes. *Gram* is sown at the same time that wheat and barley are sown, or from the middle of September to 15th November, according to latitude. *Gram* and barley are often mixed together and so sown. The gram ripens in March and April. The green pods, when opened, yield immature or green *Gram* (called *Booth* in Hindee), which in flavour somewhat resemble green peas, and are eaten both raw and cooked. This crop is very extensively cultivated. Horses and sheep are fed on it. Immense quantities are daily parched by the professional parcher, or *Bhurbhoonja*, and sold to all comers. This parched gram is very palatable, and affords a cheap and nourishing meal to day labourers and travellers. On board ship Hindee sepoys soak raw *Gram* in fresh water, and then eat the softened grain. They have to adhere to this diet all the time they are afloat, as their caste prevents them from eating food cooked on board. The *Gram*, when deprived of its shell, yields *Gram Dall* (*Chunna ka Dall* is its Hindee name), which is cooked in various ways, and is exceedingly palatable in all.

270. *Can you state how all these* Dall *crops are cultivated?*

Yes. The land after being ploughed and pulverized is ready for sowing; and all are sown broadcast, being duly weeded.

NOTE.—"The stalks of *Urhur Dall* yield the charcoal used in the Government gunpowder mills."—*J. F. Pogson.*

CHAPTER XVI.

ROOT CROPS.

Introductory observations.—Root crops cultivated in India.—The potato, the carrot, and the turnip.—The root crops which will grow perfectly, but are not cultivated by the zemindars, are cattle turnips, mangold wurzel, and arrowroot.—The sweet potato cultivated on a small scale by a particular caste.

INTRODUCTORY OBSERVATIONS.

THE Indian agriculturist, from the rulings of his caste laws and regulations, will not cultivate certain root crops. One man will die sooner than grow onions; another, by his caste prohibition, dares not grow turmeric. The sweet potato, white and red, may not be grown by any but the proper caste, who will grow yams, but will recoil from cultivating arrowroot, carrots, and potatoes. As these absurdities cannot be removed, it will rest with the European landholder to take the tabooed crops under his protection, and realize a handsome income for his audacity. Onions pay well, and arrowroot is more profitable, for, being an article of export, it is always in demand.

271. *How is potato cultivation carried on in the Himalayas?*

As soon as the snow melts, the land is ploughed up, and then laid out in furrows three feet apart. The cattle manure collected during the winter is laid in the usual manner in the furrows, and a small potato dropped in each furrow at eighteen inches apart throughout their length. The soil is raked over the potatoes and manure, and the furrow is closed. When the young crop is from six to nine inches in height, the plants are earthed up, so as to form a continuous ridge. When the plants come into bud, more soil is added on to the ridge, and the earthing up is completed. In October the entire crop is dug up, and stored in pits for next year's sale. But many zemindars prefer to sell their crop to travelling merchants, who pay on the spot, and remove the potatoes to the plains of India, where they meet with speedy and profitable sale.

NOTE.—"When the present Marquis of Tweeddale was Deputy Commissioner and Superintendent of the Hill states, from Simla to the frontiers of Thibet, he conferred a great and lasting benefit and boon on the hill population under his charge, by introducing various kinds of potatoes grown in England and Scotland. The produce of this stock was preserved, and potato cultivation annually extended, and now (1882) zemindars sixty miles north of Simla carefully cultivate the potato. The name of Lord William Montague Hay is still fondly remembered by them as the nobleman who, by introducing superior varieties of potato, brought prosperity to the doors of the population."—*J. F. Pogson.*

272. *Can you state how potatoes are prepared for planting by the British farmer?*

Yes. A potato is so cut with a sharp knife that at least two eyes are left in each part for seed. The root end of the potato should not be used for seed, but given to cattle. The rose end is the best for seed.

NOTE.—"Potato sets should always be large, and if planted whole the root end should be cut off. The writer has cultivated potatoes for twenty years, and found the best result was obtained when whole potatoes were planted."—*J. F. Pogson.*

273. *How are sets of potatoes planted?*

The furrow being manured, a set is put down at every 18 inches and covered with soil; the rest of the process being as for whole potatoes.

274. *What sort of soil is best adapted for the potato?*

A light, naturally dry, rich soil is best adapted for the potato.

275. *How does the zemindar in the plains of India cultivate the potato?*

He ploughs and prepares the land as soon as the rains are over, and plants whole or cut potatoes in furrows, and proceeds as the hill zemindar does.

NOTE.—"In the plains the potato is a winter crop, in the Himalayas a spring crop; and the main or mature crop being dug up in October, the zemindar in the plains cannot use hill seed potatoes. Hence the necessity for obtaining seed potatoes from Australia and New Zealand."—*J. F. Pogson.*

276. *Can you state which is the most profitable way of cultivating the potato?*

Yes. The German mode of culture gives the heaviest crops, which fully compensates the grower for the additional labour incurred.

The plan is as follows, viz.: Pegs are put down at every four feet, and a circle three feet in diameter described on the ground; the peg representing the centre of the circle. The soil within the circle is excavated to the depth of one foot, and in the centre of the excavation a hole six inches in diameter and nine inches in depth is dug. This hole is subsequently filled with broken bricks, *Kunkur*, or fragments of pottery, and a tile placed over its surface. The hole so prepared acts as a subsoil drain, when the super-imposed soil contains more water than is needed.

The soil taken out of the excavation is broken up, pulverized, and mixed with a liberal supply of mineral manure and good farmyard manure, after which the enriched soil is filled into the excavation and trodden down. A single potato of large size is planted in the centre of the circle, and when the plant is from six to nine inches in height it is earthed up in the usual manner. The crops should not be dug up till the haulms wither, and the yield of large potatoes will at once surprise and satisfy the grower.

NOTE.—"This plan answers admirably in the Himalayas, where all choice varieties of imported potatoes may be most successfully grown, including all the early varieties, which in Simla command very high prices.

"Potatoes may be raised from seed, and Messrs. Sutton and Sons, Reading, can supply all varieties. But the first year's produce is diminutive, and these very small potatoes when sown give a crop of small-sized potatoes of the size of a walnut. These when sown produce full-sized potatoes. Thus it takes three years to arrive at the perfected crop. Under these circumstances the better plan is to import seed potatoes during the winter months, and to sow them early in spring. The selection should be left to Sutton and Sons."—*J. F. Pogson.*

277. *Can you state what mineral matters are present in potato ashes?*

Yes. 100 lbs. of potato tuber ashes contain of potash and soda 63 lbs., of lime 2 lbs., of magnesia 5 lbs., of oxide of iron ½ lb., of phosphoric acid 18 lbs., of sulphuric acid 4 lbs., of chlorine 6 lbs., and of silica 1½ lb.

NOTE.—"Potatoes contain of water 75, of skin 3, of starch 16, of gluten 2, of fat ⅓, and of ash 1, in each 100 parts."—*Johnstone.*

278. *How does the zemindar cultivate the carrot and the turnip?*

After the land is prepared and manured the seed of each is sown broadcast, separately. The carrots when full-grown are sold as horse food, and the turnips as a culinary vegetable.

279. *Can mangold wurzel be successfully grown in Upper India, including the Himalayas?*

Yes. But it is unknown to the zemindars, and not likely to be cultivated by them, as they do not raise any root crops as cattle food. The seed of this root may always be obtained from the Government Botanical Gardens at Saharunpore.

NOTE.—"The writer has successfully grown this root crop at Simla, and he has seen it growing most vigorously in the public garden at Umballa."—*J. F. Pogson.*

280. *How is arrowroot cultivated in India, and is it grown by the zemindars?*

The zemindars do not grow arrowroot (vel *Marauta*

LAING'S SWEDE TURNIP.

NORBITON GIANT MANGOLD WURZEL.

Arundinacea), as the majority know not of its existence. They would cultivate this root if supplied with tubers for sowing. Arrowroot will not grow at high elevations above the sea-level in the Himalayas; but will grow in the valley of the Sutledge and the other Punjab rivers, in all localities where the mangoe[1] tree thrives.

Where cultivated by and for Europeans the tubers are planted in rows between two and three feet apart. Single tubers are put down at every eighteen inches in manured ground. The tubers may be sown late in spring, or just before the rains commence, being irrigated when necessary. The crop of roots is dug up in September—October, and made into marketable arrowroot in the usual manner. The crop is a decidedly profitable one.

NOTE.—"Heavy crops of tubers would be secured if the arrowroot was cultivated in the same way that potatoes are cultivated in Germany, as already explained." —*J. F. Pogson.*

281. *How is the sweet potato cultivated by the zemindars?*

The ground is marked off into plots from four to six feet square. The soil is excavated to the depth of four to six inches and piled outside. The ground inside the square is dug up and well manured, the tubers being sown singly at every eighteen inches. The crop of red and white sweet potatoes are dug up in October—November, and meet with ready and rapid sale. Hindee name, *Suk Kur Kund;* botanical name, *Convolvulus Batatas.*

[1] It does so at Kaepoo, on the banks of the Sutledge river, about seven miles distant (downhill) from Kotgurh.—*J. F. P.*

CHAPTER XVII.

COLD SPICE CROPS.

Ginger.—Turmeric.—Onions.—Garlic.—Coriander Seed.—Red pepper.

282. *How is ginger cultivated, and in what part of Upper India does it yield the best crops?*

The zemindars of the first range of the Himalayas as a rule cultivate ginger, and do so most successfully. The richest soil is selected and highly manured with cattle dung and decomposed leaves and litter from the cattle shed or house.

The ground being ready, in spring the rhizomes of seed ginger are taken out of the sheltered pit in which they have been preserved since the month of October of the past year, and as most have sprouted, the rhizomes are cut into pieces, each having a sprout. These pieces are sown singly, at intervals of eighteen inches, and, the sowing being over, the plot of land is irrigated. The leaves soon appear, and the plants grow apace, being irrigated as often as necessary. The crop is dug up in October. A portion is sold as green ginger, another portion is kept for seed, and the remainder, after being deprived of its outer skin and adherent soil, is dried in the sun for export to the plains of India as dry ginger, called *Soonth* in Hindee, whilst green ginger is called *Udruck.*

Ginger cultivation does not answer in the plains, the heat being too great. From 2,000 to 3,500 feet above the sea-level suits this valuable plant best. The dry ginger of the hills is largely exported to Southern India, and this trade has gone on from time immemorial. There are two kinds of ginger grown in the hills, the stringy and the stringless, which is reckoned the best and commands the highest price.

NOTE.—"A small plot of land under ginger enables the zemindar of the locality to pay his rent by the sale of this crop. The ginger of Dehra Doon is of the best quality, being stringless, and therefore when in the green state well adapted for making ginger preserve, such as China sends to the Indian market."—*J. F. Pogson.*

283. *Can you give a description of the Ginger rhizome, and state its botanical name?*

Yes. The botanical name is *Zingiber officinale*, being derived from its Arabic name, *Zinjeebeer*. As regards description, "The recent root is tuberous, ash-white externally; rugose, digitate, rampant, fleshy, yellowish interiorly, and covered with thin pellicles; when dried it is in fragments of about two inches long, branched, rather dichotomous, compressed at both sides, knotty, somoth, grey or ashy, with oval knots, covered by a wrinkled epidermis, parenchyma white, with ready fracture, and rather resinous appearance."—*O'Shaughnessy.*

NOTE.—"In India green ginger is daily used as a culinary condiment. The dry ginger in powder is used as a spice, and when boiling-hot water is poured over the powder the resulting tea is very warming, soothing, and comforting."—*J. F. Pogson.*

TURMERIC.

284. *Can you give the botanical and Hindee names for turmeric, and describe the substance so called?*

Yes. The botanical name is *Curcuma longa*, and the Hindee name is *Huldee*. As regards description, "The tubers are deep orange inside, bitter, and aromatic. The colouring matter of the dried root is bright yellow, soluble in alcohol and water, and changed to a deep red by alkalies. White paper dyed by an alcoholic tincture of turmeric is a very sensitive test for alkalies."—*O'Shaughnessy*.

NOTE.—"Throughout Hindoostan the boiled and then sun-dried root of the turmeric, called *Huldee*, is used as a culinary condiment. The far-famed curries of India derive their yellow colour and flavour from the use of this root. The turmeric is said to act as a deadly poison to alligators and crocodiles, which, if true, might be taken advantage of, and their destruction systematically undertaken. The powdered root, wrapped in paper, could be placed within a piece of meat, and the bait, if swallowed, would very soon bring this question to a practical issue." —*J. F. Pogson.*

285. *Can you state how the turmeric plant is cultivated?*

Yes. In the hills it is cultivated in the same way that ginger is grown, and sown at the same season. In the plains of India only certain castes may grow this crop, which is put down when the rains commence, and the roots are dug up in October. The tubers for next year's sowing are preserved in pits under shelter. The hill zemindar has no objection to turmeric culture.

ONIONS.

Introductory Observations.

The onion is indigenous to the Himalayas, where two varieties of wild but perfectly edible onions are met with. Of these the *Allium leptophyllum*, called *Bhoonna* by the Kotgurh zemindars, somewhat resembles the leek, and is improved by cultivation. The other variety grows in Kunawur, some twelve days' march north of Kotgurh, in the valley of the Sutledge river. This variety produces a regular bulb like the cultivated onion. Fifty live plants were sent to the writer from Kilba, in Kunawur, which were put down on arrival and made good growth, but as the writer left Kotgurh in October, 1881, the result of this experiment is unknown. All varieties of imported onions [1] thrive in Simla and Kotgurh, fifty miles to its north. In the plains of India the common red onion is the variety grown, excepting at Patna, in Behar, where a very fine white or silver skin variety of onion has been cultivated for over forty years. The stock or seed, it is said, came from Spain, and having been acclimatized, has given to Patna its famous onion. Now, reasoning from analogy, the very superior kinds of onion raised in the United Kingdom, if introduced and acclimatized in the Simla Hills (the Mussoorie Hills have neither the soil nor the climate needed), would yield seed which would answer perfectly in the plains.

The value of the onion as a highly nourishing esculent must sooner or later become known, and when the Brahmin votes in favour of the onion, millions of

[1] The botanical name of the onion is *Allium Cepa;* the Hindee name is *Peeaux.*

Hindoos will grow and eat it. The cultivation of superior descriptions of imported onion for regular issue to European troops is a desideratum. But until seed farms are established in suitable localities in the northern hill stations, this wholesome addition to the ration of the British soldier cannot take place.

286. *Can you state how the onion is cultivated?*

Yes. The seeds are sown in a manured seed bed, and watered every third day. When the seedlings are about six inches in height, they have to be transplanted into properly manured beds, where they will come to maturity. The seedlings should be planted in rows, six inches between plants and one foot between rows, to admit of weeding.

The manure used should be rich in phosphorus, sulphur, potash, and iron. Bone-dust, saltpetre, and sulphate of iron supply the elements needed.

NOTE.—In reference to the onion, Professor Johnstone observes: "It is interesting, therefore, to know that, in addition to the peculiar flavour which first recommends it, the dry substance of the onion is remarkably rich in flesh-formers. According to my analyses, the dried onion root contains from twenty-five to thirty per cent. of gluten."

287. *If climate and soil suited, would it pay the European tea planter to grow onions as a field crop?*

Yes. Onions yield a heavy crop, and pay well. The variety known as the "True Early Round Yellow Danvers" onion has given a yield of 1,100 bushels per acre. The bushel of onions weighs 57 lbs.

NOTE.—"Four pounds of seed is the usual quantity

needed per acre. When the onions are ripe, pull and dry them thoroughly before storing. Onions of all varieties may be most profitably cultivated at all stations where European troops are cantoned. A reference to Mr. James J. H. Gregory's catalogue will enable the reader to make a selection of onion seed; and as this well-known American seedsman has published a practical work on 'onion raising,' it should be secured.

"The onion seed, five kinds, supplied to the writer, by Sutton and Sons in 1879, produced a crop of onions which quite astonished the European and native residents of Kotgurh."—*J. F. Pogson.*

GARLIC.

288. *Can you give the botanical and Hindee names for the garlic?*

Yes. The botanical name is *Allium sativum*, and the Hindee name is *Lussun.*

289. *How is the garlic cultivated by the zemindars of India?*

The ground having been prepared as for onions, the heads of garlic are broken up, and a single clove of garlic put down in rows, at every six inches apart. The space between rows to be one foot. The planting varies according to the seasons in the hills and plains. When the bulbs are ripe (October in the hills, November, December in the plains) they are pulled up, sun-dried, and sold to the buneeah, who, as a rule, supplies the population with all cold spices.

NOTE.—"The peculiar smell of garlic is due to its oil.

The strength of its odour may be judged from the fact that, powerfully smelling as garlic is, from thirty to forty pounds of it are required to yield one ounce of the oil."—*Johnstone.*

CORIANDER SEED.

290. *Can you give the botanical and Hindee name of the plant which produces coriander seed?*

Yes. Its botanical name is *Coriandrum sativum*, and its Hindee name *Dhunneeah.*

291. *How is this plant cultivated?*

All Hindoos may grow it. The land being prepared, the seed is sown broadcast. The green leaves are considered a treat by the natives, but have an offensive smell and taste, repugnant to the European palate. The seeds ripen in the cold months, and are speedily sold to the buneeah.

CAPSICUM, OR RED PEPPER.

292. *Can you give the botanical and Hindee name of the plant which produces red pepper?*

Yes. Its botanical name is *Capsicum annuum*, and Hindee name *Laull Mirrich*, and when green, *Hurra Mirrich.* The bird's-eye pepper is called *Dhan Mirrich.*

293. *Can you state how the zemindar cultivates this plant?*

Yes. The seeds are sown in a seed bed, and when from four to five inches in height the seedlings are transplanted into prepared plots of land, and put down at intervals of eighteen inches. They are watered when necessary, and the pepper pods are ripe in October, when they are gathered, sun-dried, and sold as before.

294. *Can you state what substances are present in the ripe capsicum?*

Yes. The chemist, Braconnot, has shown that in one hundred parts the capsicum contains of

Starch	9·00
Acrid Oil	1·90
Wax and Red Colouring Matter	0·90
Gum	6·00
Azotized Matter...	5·00
Salts of Potash	6·00
Lignin	67·80
Muriate and Phosphate of Potash...	3·40
	100·00

NOTE.—"The colouring matter is due to iron. Bone manure and potash are needed by the plant."—*J. F. Pogson.*

CHAPTER XVIII.

FODDER PLANTS.

Introductory observations.—The fodder plants cultivated by the British farmer named.—Why all do not succeed in the plains of India.—Fodder plants which thrive in the hills, but not in the plains.—The buffalo grass, or *Reana Luxurians.*—How cultivated.—Its value.

INTRODUCTORY OBSERVATIONS.

THE zemindars of India do not cultivate any crops as a fodder plant, and if they can be induced to do so, it will only be for the express purpose of selling the produce to Europeans, and the Government executive commissariat officer, charged with the duty of providing dry and green fodder for various animals, from the elephant, down to the battery mules and stud donkeys. All the European fodder plants will thrive in the Himalayas, but the heat of the plains is too much for the clovers. Lucerne will succeed if duly irrigated. But the various kinds of pasture grasses so carefully cultivated in the United Kingdom cannot withstand the great heat of the Indian sun, which scorches up all grass not under regular irrigation.

295. *Can you name the fodder plants cultivated by the British farmer?*

Yes. He cultivates red and white clover, Lucerne, and Italian rye grass, and, in addition, various kinds of pasture grasses.

296. *Which of these will answer in the plains of India?*

Lucerne, when sown in spring and regularly irrigated, keeps green throughout the hottest weather. The heat is too great for the others.

NOTE.—"The writer transplanted Lucerne at Umballa early in spring. The plants grew vigorously, and by August were three feet in height. All the transplanted plants flowered and matured seed, whilst the plants left *in situ*, though over three years old, flowered, but never bore seed. At Kotgurh the Lucerne withstands the cold, but does not flower owing to the soil being most deficient in lime.

"211½ lbs. of Lucerne ash contain 107½ lbs. of lime. Hence it follows that the soil should be liberally limed if heavy crops of Lucerne are desired."—*J. F. Pogson.*

297. *Can you state how the buffalo grass, or* Reana luxurians *of botanists, should be cultivated?*

Yes. The seed should be sown in spring, in seed beds. Two seeds should be put in each dibble, the ground being previously limed and manured. The space between dibbles should be four inches. The third day after sowing, water the seed bed, and so continue every other day till the seedlings appear; after this, water once a day. When the seedlings are from six to nine inches in height, they are ready for transplanting. To transplant, proceed as follows: At every four or four and a half feet put down a peg, and with it as a centre, describe on the soil's surface a circle, eighteen inches in diameter. Dig up the soil within the circumference of the circle, apply those to two chittacks of *Chunam*, previously mixed with one hundred grains of saltpetre, and add a double handful of cattle manure. In this way prepare the land (which may be waste) intended to be placed under this wondrous fodder plant. Then when everything is ready commence transplanting, and take the seedlings out of the seed bed, with a ball of earth attached to each, and carefully put down in the centre of each circle, and water daily until the plants make growth, when they may be watered at discretion. As soon as the rains set in, the crop will make rapid growth, and in due time flower and mature seed.

298. *How should this fodder plant be cut for use, and how preserved?*

The plant will throw out numerous stalks from three to six feet in length. When needed for use as green fodder, six or more stalks with leaves attached should be cut from each plant, and so given to the quadruped in need of it. This cutting to be continued throughout the rains. When fair weather sets in, with the exception of some plants reserved for seed, the entire crop should be cut down and sun-dried. This gigantic variety of hay should then be stacked.

NOTE.—"From five to six feet of space should be left between each row of piants. When the crop is cut for haymaking, it would be advisable to make the produce of every two or three sets of plants into bundles, and to stand them on end, so that the sun and air may act together and induce rapid drying. Should it rain, the hay would not be sodden, which would be the case if the cut grass was laid on the soil's surface and so allowed to dry. When cutting, at least one foot of the stalks should be left on the roots, which may survive a mild winter, but do not stand the frost. If an ample supply of water be available, the plants may be cut for green fodder as soon as the leaves are from four to six feet in length. There is no fodder crop as yet known in India which can equal the *Reana luxurians*, and its extended cultivation is strongly recommended. It thrives in the Himalayas from April to October."—*J. F. Pogson.*

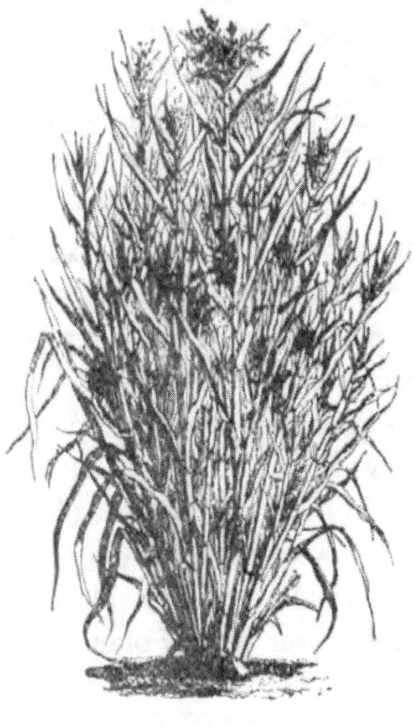

BUFFALO GRASS (*Reana luxurians*).

CHAPTER XIX.

WATER-NUT CROPS.

Introductory observations.—Botanical and Indian names of the water-nut.—How cultivated.—The nut harvest, how collected. —The sun-dried nuts used for food.—The kernel of the nuts eaten green or uncooked.—Method of boiling and roasting the nuts.—The kernel of the sun-dried nuts, when reduced to powder, used in making porridge and sundry bazar sweetmeats. This nut-meal being very nutritious should be used in place of arrowroot by the sick and convalescents.—The water-nut cultivated in the United States of America and in Australia.—The cultivation of the water-nut in the Madras Presidency, as a preventive against famine, advocated.—The nuts being at all times saleable, a considerable annual income could be so derived if the freshwater lakes extant in the Madras Presidency were to be systematically put under water-nut culture as a measure of State.

INTRODUCTORY OBSERVATIONS.

THE Chinese have cultivated the water-nut for some three thousand five hundred years at least, and recent discoveries made in Europe tend to prove that the prehistoric races, who built their dwellings on piles driven into the beds of freshwater lakes, also cultivated this nut as an article of food. In India the water-nut was cultivated by the race of men whose language was Sanscrit, and who may have brought down the seed nuts from Cashmere, whose grand freshwater lakes are to this day made to produce an annual harvest of water-nuts, the sale of which is said to yield an income of some £12,000, the quantity obtained being sufficient to feed 30,000 persons for five months. And as this water-nut will grow in lakes and tanks all over Hindoostan, a very moderate outlay would suffice to supply seed nuts to the agriculturists of all districts who may possess tanks and natural reservoirs of fresh water, but are not acquainted with the nut and its very high value as a palatable, sustaining, and nourishing food.

299. *Can you give the botanical and Indian names of the water-nut, and state if any variety exists?*

Yes. The botanical name of the Indian water-nut is *Trapa bispinosa*, and its Hindee name is *Singhara*. The variety met with in Europe is called the *Trapa bicornis*. The water-nut of India has two strong sharp-pointed thorns, placed one at each end of the shell which covers the kernel of the nut; whilst in the European variety a short curved horn occupies the place of the thorn of the Indian nut, which is unknown in Europe and is considerably larger than the *Trapa bicornis*.

300. *Can you state how the Indian water-nut is cultivated?*

Yes. Where the water is deep and permanent the nuts once sown in the bed of the lake, tank, or pond strike root and send a leafless stem to the surface, which developes leaves, and in due time flowers appear, and the very curious water-nut is formed and floats on the water's surface. This shows that in deep water the nut plant becomes a perennial.

The tanks and natural reservoirs or depressions of the

soil which get filled with water during the rains, are, in most instances, quite dry during the severe and prolonged hot weather of India. As soon as the rains fairly set in the zemindars dig up the seed water-nuts which have been preserved under shelter (just as seed ginger is kept underground till wanted), and proceed to sow them in the bed of the tank at intervals of three or four feet, and in this way plant the bed of the entire tank, which gradually fills as the rains advance. In due time the nuts germinate, the stem finds its way to the surface, the leaves appear, then the flowers and nuts. The harvest follows, and when the hot weather once more comes round, the plants wither and die, and thus ends the water-nut crop, when, from want of water, the plant becomes an annual.

301. *Can you state how the water-nut harvest is gathered?*

Yes. Two methods are adopted. First, when the water is not more than waist-deep, the zemindars or owners being provided with common pottery vessels called *Handees*, enter the water, and, proceeding from plant to plant, break off all mature water-nuts and put them inside of the pottery vessel which floats by their side, and so proceed till it is nearly filled, when the man places the vessel on his head and walks out of the water. In this way, day after day, the zemindar visits his water-nut plantation till the season ends, and before it does so he secures his supply of seed nuts for the coming year. The second plan is a far more dignified affair, boats and rafts made of pottery vessels being used to collect the harvest of ripe nuts. The nut-gatherer sits on his raft and paddles himself from one plant to another, and the nuts as broken off are thrown into the pots, until no more are wanted. The man in the boat proceeds in the same way, only he gathers on the large scale, and goes home when he has secured a boat-load.

302. *Can you explain how the water-nuts are preserved for future use?*

Yes. The kernel of the nut is covered with a thick, tough skin, and by placing the nuts in the sun this dries and preserves the kernel for an indefinite period. When required for use the shell is cracked, the kernel removed, and it may be eaten uncooked or boiled. The dry kernels when ground yield a meal which may be made into porridge or gruel, and when mixed with sugar and boiled butter into sundry sweetmeats. This meal is very nutritious and easy of digestion, hence of superior value to arrowroot, cornflour, &c. As sustaining food for the sick it has everything in its favour, but few know of its existence.

NOTE.—"The kernels of the fresh-gathered nuts may be, and are, largely eaten raw, as extracted from the shells. The fresh nuts are also boiled in their shells, and when cold the kernels are taken out by the purchaser and eaten at his leisure and pleasure. These boiled kernels have the flavour of boiled chestnuts, and are nearly as good eating as boiled potatoes. The Chinese roast the water-nut, and no doubt use the shells, or shell-ash, as manure. The Hindoo throws them away."—*J. F. Pogson.*

303. *Can you state how the water-nut is cultivated in the United States of America?*

Yes. Under the name of "water chestnut," the *Trapa bicornis* is planted and grown in muddy brooks. This aquatic plant is considered very beautiful by the

Americans. It has only been recently introduced, and in place of being eaten, the ladies consider that, "it makes an elegant little ornament for the parlour table."

304. *Can you state how the Indian water-nut came to be introduced into Australia?*

Yes. A correspondent of the Agri-Horticultural Society of India earnestly suggested that all freshwater lakes, tanks, and *Jheels* to be met with in the Madras Presidency, which includes Southern India (the scene of the last terrible famine), should be systematically planted with the water-nut (*Trapa bispinosa*) of the Bengal Presidency. The Baron Ferdinand von Müller, of Melbourne, read this communication, and, appreciating its practical value, took steps to obtain the seed nuts from India, which were duly sent him from Calcutta; and now Australia possesses both kinds of water-nuts, with which the freshwater lakes of Australia will gradually be planted, and an inexhaustible supply of wholesome food secured.

NOTE.—"For a country liable to periodical famines, and yet possessing freshwater lakes, spoken of as inland seas of fresh water, the *Trapa bispinosa*, or *Singhara*, is not only a stand-by, but the most valuable of food-producing plants. The unshelled, sun-dried nuts may be stored in pits and kept for years to meet all contingencies, or the stored supply may be sold as soon as the grain crop is in ear. The British manufacturer of farinaceous food would be only too glad to meet with a natural product which may be turned to profitable account in various ways."—*J. F. Pogson.*

305. *Can you state why the water-nut has not been planted in the freshwater lakes of Madras?*

Yes. These lakes as well as all *Jheels* are Government property, and as such no private person dare cultivate any portion of their beds with the water-nut. The Government of Madras can alone take the initiative in this matter.

CHAPTER XX.

GROUND-NUT CROPS.

Introductory observations.—How grown in America.—Description of ground-nut.—The oil a substitute for olive.

INTRODUCTORY OBSERVATIONS.

THE ground-nut, called in America the pea-nut, is a native of China, and is called *Cheena Boodaum* (*Anglicè*, Chinese almond), in Bengal proper. The nut is now cultivated in various parts of the world, and will thrive wherever the potato does so. *Mode of culture.*—It is a valuable crop, and only requires suitable soil and mineral manure to yield 2,500 to 3,000 lbs. weight of nuts per acre. A kind of chocolate is made from the pea-nut, and when pressed, the nuts yield a valuable edible oil.

306. *Can you give the botanical and Hindee names of the ground-nut?*

Yes. The botanical name is *Arachis hypogæa*, and the Hindee name is *Moong Phullee.*

307. *Can you state how this nut is cultivated?*

Yes. The land being prepared and laid out in ridges, two feet apart, a tablespoonful of *Chunam* and bone-dust mixed together is to be placed on top of the ridge, at intervals of eighteen inches, and dug into the soil. A single entire nut-pod is to be planted in the centre of the manured soil, at the depth of one and a half inches, and covered over with manured soil. In a few days the young plants will appear, and grow vigorously ; in due time pretty orange-coloured flowers will bloom, followed by the pods, which as they increase in size force themselves underground, where they come to maturity.

NOTE.—"The ground-nut was thus cultivated at Simla by the writer. The nuts were planted in April, and dug up in October, as soon as the frost appeared. The nuts so raised were nearly twice the size of those sown, and the kernels were larger, and had a delicious flavour."—*J. F. Pogson.*

308. *Can you state how the ground-nut is cultivated in America?*

Yes. The American farmer selects a calcareous soil, which he ploughs and prepares for this crop, as experience has taught him that the pea-nut will not fruit except on a calcareous soil. If sown on land poor in lime, the pods do not fill; and the use of bone manure is essential to success.

The American farmer does not sow whole pods, but, previous to planting, causes them to be carefully shelled, and every faulty bean thrown out; not even the membrane enclosing the seed should be ruptured; and on this plan it takes about two bushels of pea-nuts in the pod to plant an acre. The bines make good hay, and yield one ton of it to the acre. The seeds are sown singly at intervals of eighteen inches in furrows, as for potato culture.

NOTE.—" In the State of Virginia, U.S., the bushel of

pea-nuts weighs 22 lbs., and in Carolina, U.S., 28 lbs. In good soil one hundred bushels per acre are produced; but in poor soil the yield is from 25 to 30 bushels per acre.

"The above information is taken from a recently published American work."—*J. F. Pogson.*

309. *Can you state why amateur cultivators of the ground-nut in Upper India do not always obtain satisfactory results?*

Yes. Be they Europeans or natives, they know little about mineral manures, and care less; and as calcareous soils are not common in the plains of India, the crop when manured with cow-dung ensures a failure. If by chance the soil is rich in lime (of which the cultivator knows nothing), a success more or less partial follows, and then the man who fails puts down such success to chance, not being aware that without lime, bone-dust, or, better than all, fossil phosphate of lime, the ground-nut will not thrive.

310. *Can you give a description of the ground-nut, plant and nut?*

Yes. The flowers have the pea form, on the diminutive scale, and their colour is orange; the leaves somewhat resemble the pea leaf, and the stems or haulms, instead of climbing, creep along the surface of the soil, and the ridge form of culture having been selected, the pods as formed imbed themselves in it, and produce full growth underground. The fruit (legume) is oval oblong, coriaceous, with a network of veins, containing two oval, somewhat flattened seeds, covered with a brown arillus. These seeds contain, according to Professor O'Shaughnessy, an abundance of fixed oil, resinous matter, lignin, crystallizable sugar, gum, colouring matter, sulphur, starch, and saline substances.

NOTE.—"It may interest the British farmer to know that the ground-nut, being an hardy annual, may be successfully grown by him. The roasted nuts are not only exceedingly palatable, but constitute a true and rich nourishing food. The nuts as dug up have only to be pressed to yield a valuable oil (of which presently), and the oil-cake affords excellent fattening food for cattle."—*J. F. Pogson.*

311. *Has the ground-nut oil any particular properties which distinguish it from other bland oils?*

Yes. Arachis oil, when pure, is limpid and nearly colourless; it is insoluble in alcohol, but soluble in ether and turpentine. It has, while fresh, a faint smell, and but little taste. The oil, according to Professor O'Shaughnessy, is calculated to serve as a perfect and very cheap substitute for olive oil, for pharmaceutical purposes. When used as a lamp oil, it gives a very full bright light, with a clear flame and little smoke. Its specific gravity is 0·9163.

312. *What quantity of oil do the kernels of the ground-nut yield?*

To put this matter to the test, 1,950 parts of seed separated from their coverings were blanched, and gave 1,405 parts of kernels, from which by cold pressure 703 parts of oil were obtained.

NOTE.—"The above information is official, and shows that the ground-nut is of considerable value an an oil producer."—*J. F. Pogson.*

313. *How should the kernels of the ground-nut be treated, in order to yield a beverage somewhat like chocolate?*

The kernels should be roasted, just as is done with green coffee-berries, care being taken that they are neither burned or charred. When done, let them cool, and then reduce to powder, with some sugar added. An ounce of this compound, boiled up with a teacup of water, will yield the beverage desired, to which sugar and milk may be added at discretion.

NOTE.—When *Cacao theobroma,* pods of Indian growth, are easily obtainable, it will be quite feasible for all whom it may concern to make their own chocolate. The kernels of the cacao-nut, in certain proportions, being roasted, reduced to paste, and mixed with a similar paste made with the lightly roasted kernels of the ground-nut, will yield, when pressed into moulds, a perfect cake chocolate, within the means of all classes, whereas the present price of imported cake chocolate, which costs in the shop from Rs.2 to Rs.2 annas 8 per pound, is prohibitive.

CHAPTER XXI.

THE RUSH-NUT, *vel* CHUFAS.

Botanical name of the *Chufas.*—It has no Hindee name.—The rhizomes described.—Their composition and flavour.—How cultivated in America.—How cultivated in Spain, and the yield per acre.—The value of the nut as a fattening cattle food.—*Chufas* coffee, how made.—The nut has recently been introduced into India by the Agri-Horticultural Society of India. —The extension of the cultivation a desideratum, but not feasible, owing to the want of seed farms.

314. *What is the botanical name of the Rush-nut, or* Chufas?

Botanists call it the *Cyperus esculentus*, Anglicè, esculent *Cyperus*. It has no Hindee name.

315. *Can you give a description of this peculiar nut?*

Yes. The rhizomes are composed of thin fibrils, to the extremities of which are attached round or oblong bulbs, externally brown, white within, substance white, tender, and farinaceous; odour slightly camphorate.

316. *What substances are present in these nuts, and what flavour have they?*

"The bulbs on analysis afford starch, fixed oil, sugar, albumen, gum, salts of malic acid and tannic acid, and oxide of iron." As regards flavour, the nuts "taste sweet and agreeable, rather saccharine and mucilaginous."—*O'Shaughnessy.*

NOTE.—"The dry root contains one-sixth of its weight of oil of a beautiful golden colour, nut-like smell, slightly camphorate taste; specific gravity 0·918. It deposits stearine by rest, and in its general properties resembles the other fixed oils."—*Vide* "*Bengal Dispensatory.*"

317. *How is this nut crop cultivated by the American farmers?*

In Massachusetts, the ground being duly prepared, single seed-nuts are planted in lines, at intervals of one foot between nuts; the space between the lines being eighteen inches. The plant is very prolific, a single nut yielding from two to four hundred. These nuts very closely resemble in sweetness and richness of flavour a cocoa-nut.

318. *How is this nut-crop cultivated in Spain?*

The ground being prepared is laid out in drills, from 30 to 36 inches apart, and single nuts are sown at every 12 inches. In Spain, the yield varies from 200 to 500 bushels per acre. The tubers when ripe and dry are sweet to the taste, somewhat resembling almonds; it is owing to this that the fanciful name of "earth almond" has been given to the *Cyperus esculentus.*

NOTE.—"The soil of Spain is rich in lime, which the ordinary soil of the Bengal Presidency is not. Hence it follows that the same mineral manures will be needed for this nut crop as is requisite for the *Arachis hypogæa*, or ground-nut. The immense yield indicated shows the value of this crop under suitable culture; and the composition of the edible portion of the nuts points out what should be used as manure."—*J. F. Pogson.*

319. *What quantity of seed-nuts are needed to plant an acre of land?*

One bushel is the quantity usually planted per acre. The single nuts should be planted two inches in depth, in a dibble, and covered with fine soil, properly enriched with mineral manure. The growing crop is to be carefully weeded, and the land kept clear of weeds. Water when necessary.

NOTE.—"The seed-nuts supplied to the writer failed to germinate. In the Himalayas they would have to be sown in March to April, but the proper time of sowing in the plains has yet to be decided. September would most probably be the best month for sowing."—*J. F. Pogson.*

320. *How are Chufas coffee, and Chufas chocolate prepared from these nuts?*

They are roasted like ground-nuts, and if highly roasted and powdered, will, on being boiled in water and strained, yield the coffee; whilst if lightly roasted, and made into a paste, and boiled with pure milk, or milk and water, the beverage obtained will be *Chufas* chocolate.

NOTE.—"As these nuts are rich in albumen, gum, oil, saline matters, and oxide of iron, the chocolate made from them would be very nutritious; and perhaps if a cup of decoction of coffee, or a spoonful of essence of coffee, was added to the preparation, the compound would then contain all that was needed, and be as nourishing as chocolate; the cafeine supplying the missing component. But even without it a wholesome and nutritious warm beverage would be available for millions."—*J. F. Pogson.*

321. *Can you state when the cultivation of the* Cyperus Esculentus *was brought to the notice of the Indian authorities, and when the actual introduction took place without their aid?*

Yes. In 1837 a committee was, by order of the Government of India, appointed to carry out a scientific scheme for giving India a *Materia Medica.* The work was published in November, 1841, and at page 628 the *Esculent Cyperus* was thus brought to official notice. The cultivation of the plant deserves attention for its considerable alimentary value.

A terrible famine had devastated the North-Western Provinces in 1837-38. Hence to have warmly advocated the cultivation of the plant named, would have been treading on dangerous ground. Therefore it was summarily disposed of, and whilst the Americans have taken full advantage of the alimentary value of the *Chufas*, it is extremely doubtful if the Indian Agricultural Department has awakened to a sense of its value. Would millions have died in India during the last famine had this prolific and easily grown nut been everywhere available?

To the Agri-Horticultural Society of India belongs the credit of having introduced this nut, and with its philanthropic members will rest the honourable duty of extending the cultivation.

NOTE.—"In America these nuts are reckoned to afford superior fattening food for pigs; and as sheep, cows, and oxen would thrive on them as well, their value as food is very considerable.

"The periodical recurrence of famine in India renders it necessary that the food supply should be increased. In the water-nut, the ground-nut, and the *Esculent Cyperus*, we possess three valuable food crops, whose regular and systematic cultivation should no longer be neglected."—*J. F. Pogson.*

CHAPTER XXII.

COTTON CROPS.

Botanical and Hindee names of the cotton plant, annual and perennial.—The varieties cultivated in different parts of the world.—The annual cotton plant described.—The plant, how cultivated.—The yield of cotton wool due to the soil being suited to the plant, as at Jullunder, and Hoshiarpore in the Punjab.—The yield of cotton wool per acre at the Cawnpore Experimental Farm stated for comparison.—Cotton crops, how manured at Cawnpore.—The remarkable results of the system.—Analysis of cotton seed.—Cotton seed oil.—Cattle fed on cotton seed.—Its Hindee name.—Proper manure for cotton.—Cultivation of the *Gossypium arboreum*, or sacred cotton tree of the Deccan, advocated.

322. *Can you give the botanical and Hindee names of the cotton plant?*

Yes. The botanical name is *Gossypium herbaceum*, and the Hindee name for all annual varieties is *Kuppaus*. The cotton wool is called *Roo-ee*, and the cotton seed, *Binowla*.

NOTE.—"The Arabic word for cotton is *Kuttun*, and the English word is no doubt derived therefrom."—*J. F. Pogson.*

323. *Can you name the varieties of cotton cultivated in different parts of the world?*

Yes. In America, ten varieties are under culture.

They are—1. Sea Island Cotton: 2. Brazilian Cotton: 3. Improved Sea Island: 4. West Indian: 5. Demerara: 6. Pernambuco: 7. New Orleans or Mexican: 8. Tennessee: 9. Peruvian: 10. Upland Georgia Cotton.

The African varieties are—1. The Helena Vine-leafed Cotton: 2. Mangrole: 3. Seychelles: 4. Fine Bourbon: 5. Ordinary Bourbon: 6. Egyptian: 7. Bamiah Cotton.

The Indian varieties are—1. The Nurma, a peculiar fine soft cotton, cultivated in parts of Rajpootana for making very superior *Puggrees*: 2. Dacca Cotton, from which the celebrated Dacca muslins were made: 3. Dharwar Cotton: 4. Malabar Cotton: 5. Daesee, or common indigenous Cotton: 6. Assam Cotton: 7. Siam Cotton: 8. Burman Cotton: 9. Chinese or Nankin Cotton, which yields cotton wool of a drab colour, or light shade of brown: 10. Sacred Cotton.

NOTE.—"The reports of missionaries who have travelled in Africa tend to show that varieties of cotton exist which are so prolific as to require bearing branches to be supported, to prevent their breaking down under the load of cotton pods. This is evidently a perennial or tree cotton, and perhaps identical with the *Gossypium arboreum* of the Deccan, which bears beautiful crimson-coloured flowers."—*J. F. Pogson.*

324. *Can you give a description of the annual cotton plant?*

Yes. The leaves are generally palmate and hairy; and the blossoms are large and of a beautiful yellow colour. The common cotton grows in the bush form, and up to the height of from three to five feet. Cotton wool is a filamentous down which invests the seeds of the plant, and these grow within a heart-shaped pod, which bursts when the seeds are ripe and discloses its contained snow-white cotton wool.

NOTE.—"Cotton is extensively cultivated in the Himalayas, where the soil and altitude above the sea-level suits the plant, which withstands very sharp frost. The hill cotton pod bursts into four divisions, each containing its quota of seed cotton. The cotton pod of the plains contains only three divisions, similarly supplied; hence the hill cotton contains more cotton wool than does the cotton pod of the plains. It grows at Hurreepore,[1] and in the adjacent valleys."—*J. F. Pogson.*

325. *How is the cotton plant cultivated by the zemindar?*

The ground being ploughed and pulverized, the cotton seed, previously smeared over with fresh cow-dung, which prevents the seeds sticking together, is sown broadcast and harrowed in. If water is available the sown field is irrigated. In due time the plants appear and are thinned out at discretion. Beyond weeding, little care or attention is paid to the growing cotton crop. The pods begin to burst in October, and they are daily gathered till the crop is harvested. The cotton sticks are next pulled up to be used as fuel, and the land is free for future use. In localities where irrigation is not available the zemindar waits till the rains begin, when the land is ploughed and sown as above stated.

326. *Can you suggest an improved method for cultivating the cotton plant?*

Yes. The seeds should be sown in a seed bed, the soil of which should be properly manured. The seeds, previously prepared by being soaked in a pickle (of

[1] Hurreepore is on the old road from Kalka to Simla.—*J. F. P.*

which presently), are to be sown in dibbles made four inches apart and about one and a half inches in depth. Two seeds are to be sown in each dibble, and the seeds covered over with fine soil, so as to fill the dibble. The bed being sown should be lightly watered, and the watering should be daily repeated.

The plants will very soon appear above ground, and are to be allowed to grow, weeds being carefully removed.

The seed cotton may be thus sown between the 15th of April and 10th of May, and by the time the rains commence the seedlings will be ready for transplanting.

The seed beds should be three feet broad and six feet in length, inside measure. The square foot of soil will contain sixteen dibbles, requiring thirty-two seeds, and as the seed bed contains eighteen square feet, 567 cotton seeds will be needed per bed, of which there will be seventeen. A path one foot wide should be made between the beds, so that the seedlings may be watered and weeded without the seed bed being entered.

NOTE.—" *To make the Pickle for Cotton Seed.*—In a gallon of water dissolve one tablespoonful of *Chunam*, and add thereto one dessertspoonful of saltpetre in powder, and the like quantity of the sulphate of iron also in powder; stir till dissolved, when the pickle is ready for use.

" *To use the Pickle.*—The pickle being made in a tub, bucket, or pottery vessel of suitable size, the quantity of cotton seed needed for the seventeen beds is to be put into the solution or pickle at sunset, the seeds being turned over by hand to be equally wetted. They are to be soaked all night, and are to be taken out of the pickle at sunrise, and spread on mats in the shade, and they are now to be sown in the seed beds as quickly as possible. If any one of these seeds be opened, it will be seen that germination has set in, and this accounts for the rapid appearance of the plants as stated."—*J. F. Pogson.*

327. *How should the land be prepared for the reception of the seedlings?*

The acre of land being ploughed and pulverized is to be laid out in drills about three inches deep, and at every three feet of length a handful of mineral manure is to be dropped, which another man is to mix and dig into the soil. The space between drills is to be three feet, and the number of seedlings to be put down will be one to the square yard. As two seeds were sown in each dibble, two seedlings will, as a rule, be found growing together, and care must be taken when transplanting them not to break the ball of earth in which they are growing. The ball of earth, with its single or double seedling, as the case may be, is to be carefully put down at each manured spot, and as the rains will have set in watering will not be needed, but if otherwise irrigation will be necessary. Hence transplanting operations should be postponed till the rains commence.

NOTE.—By adopting this plan the grower obtains a start of at least one month over the man who sows his cotton seed broadcast when the rains set in. The result is obvious, and before the crop sown in the old way is half grown, the cotton cultivated as stated will be in flower, and the pods will begin to ripen as soon as the rains are over. The writer adopted this plan of cotton culture at Umballa in 1873 with perfect success. By allowing one square yard of ground to each single or double seedling, ample room is afforded for horizontal branch growth; and as no thinning out is necessary, waste of seed and destruction of plants is avoided.

328. *Has the yield of cotton wool anything to do with the nature of the soil on which the cotton plant has been grown?*

Yes. A great deal depends on the soil, which if calcareous will yield heavy crops of cotton wool, the production being increased if the plants be suitably manured. A light sandy soil or a stiff clay soil are equally unsuited to cotton culture, and the soil of Cawnpore is an example. It is a simple waste of time and money to put such land under cotton; for, though the plant will grow and bear cotton pods, the yield with suitable manure will be unsatisfactory, whilst with unsuitable manure it will be insignificant. All soils which produce good crops of sugar-cane will also produce good cotton crops. In the Punjab such soils are common, and indicate the districts in which cotton cultivation should be fostered.

NOTE.—"To show that no amount of manuring will make unsuitable soil produce a heavy yield of cotton wool, the reader's attention is invited, first, to the extracts given beneath, and taken from the Proceedings of the Agri-Horticultural Society of India for February and April, 1882; and, second, to Mr. Bennett's report on cotton cultivation as carried on at Cawnpore.

"In February, 1882, Mr. C. S. Faddy, district engineer, Hoshiarpore, sent three samples of seed and cleaned cotton to Calcutta for report, which was very favourable.

"Mr. Faddy caused 30,380 square feet of land to be sown with cotton seed, drilled in lines four feet apart.

One maund, or 82 lbs., of bone-dust were applied, and the yield was a few seers over 10 maunds, or over 820 lbs. of seed cotton from the plot. Further on Mr. Faddy reports: 'It is quite possible with rational culture to raise 1,100, 1,200, or even 1,600 lbs. of cotton to the acre, giving when cleaned 500 to 600 lbs. of a good marketable commodity; and, with the exception of sugar-cane, it is the most profitable crop a ryot can cultivate.' Mr. Shircore, of Jullunder, sent cotton of his growing to Calcutta in April, 1882, and as the official report is too valuable to be abridged, it is given in full, to satisfy the sceptic, the capitalist, and the critical reader that the Punjab can compete with America as a cotton-producing country; the only thing needed being seed farms for producing acclimatized seed-cotton."—*J. F. Pogson.*

Read the following letter from Mr. A. A. Shircore Jullunder City, on two samples of cotton grown by him:

"I have much pleasure in sending you, per parcel post, two samples of cotton grown by me, one sample with seeds, the other without.

"The yield has averaged about 600 lbs. of cotton with seeds per acre, on a small plot of land irrigated by well water. I consider this return as normal, and the same yield can be obtained by myriads of cultivators in this and the adjoining districts of Hoshiarpore, Kappurthalla, Ludhianna. A large return could very easily be obtained on first-class soil and by good cultivation. I think it quite possible to produce 800 lbs. of uncleaned cotton in an acre of land.

"The cultivators in these parts would gladly avail themselves of producing any profitable crops, if they only knew how to go about it. They appear to be very backward in agricultural matters.

"The Government does next to nothing, and there are no European traders or residents to improve their condition.

"The cotton, you will readily see, is from American seed. I am unable to state what particular class it is styled in the trade; but I believe there are two or three varieties, judging from the pods produced. The seed was sent to me by the Government of Bombay, in one bag, without any particulars being noted.

"A large amount of cotton is grown in these parts on irrigated and unirrigated land, but it is of a very inferior description, as you will see from the enclosed sample of the indigenous kind, after comparing it with that grown by me.

"The cotton has been ginned by the usual native method. The proportion of cotton to seed is as 68 to 32 per hundred.

"It is a great pity the agricultural resources of these parts are literally wasted.

"If Government took vigorous and forced measures to improve the quality and yield of cotton, I am of opinion a very large amount, and of superior quality, could be exported to the English and European markets in a couple of years to the enrichment of the Government and ignorant peasants. Will you be kind enough to let me know the opinion of the Society on the cotton sent, and the price it will fetch in European markets?"

"Submitted the following reports thereon:

"The Jullunder cotton is much better than the samples recently sent to the Society from Hoshiarpore for report; it is much superior in staple to the generality of cotton grown in Bengal, North-West, or Central Provinces, and would spin easily into 40s. to 50s. twist. It has good strength and colour, and at Liverpool would class with Orleans and be worth at present 5⅞ or 6d. per lb. With the yield reported by Mr. Shircore, 600 or 800 lbs. per acre of cleaned cotton, it ought to pay well to cultivate, and would not require any aid from Government.

"The native cotton would class with first quality Cawnpore worth at present Rs. 16-8 per maund in Calcutta.

Both samples are inferior in colour in the cleaned state to those with the seed—probably from bad ginning.

(Signed) "S. H. ROBINSON."

"The Jullunder sample grown by Mr. Shircore is most creditable. The staple is even, soft, almost silky, good colour, and of fair strength, of too good a quality for consumption in this country (India), as the local mills spin low counts only of yarn, but for 40s.

and upwards the twist manufacturers in England would be ready buyers at fair market prices. It partakes more of the Sea Island than the Orleans type, in appearance and softness to the touch, and might realize 7d. per lb. Such a result in quality and quantity produced proves that the climate and soil are favourable, and that the cultivator possesses the needful knowledge to enable him to act independently of Government aid.

"If Mr. Shircore placed the result before the leading zemindars and cultivators, there is every inducement for them to follow his good example, and reap the full benefit from the use of good seed and improved cultivation.

"The native grown sample is of the ordinary class of fair Bengal short staple, and worth about Rs. 15-8 per maund in the Calcutta market. The difference in colour between the samples in their unginned state, or not churkaed, and the cleaned one, may, in my opinion, be guarded against in separating the seed more carefully, and not exposing the cotton to dust, dirt, and to less dirty handling.

(Signed) "W. H. COGSWELL."

These results have been attained without resorting to any improved method of culture such as adopted at the Cawnpore farm, where three fields were treated differently. Plot A was deep ploughed two months before, and again at sowing-time; plot B was deep ploughed at sowing-time only; and plot C was country ploughed twice at sowing-time, just as native cultivators do. The results were as follows per acre: Plot A gave of cleaned cotton, 252 lbs.; plot B, 153 lbs.; plot C, 177 lbs. An acre of light loam sown with New Orleans cotton gave 109¾ lbs. of clean cotton. An acre of upland Georgian cotton gave 92¼ lbs. of clean cotton. An acre of heavy loam sown with New Orleans cotton gave 263 lbs. of clean cotton; whilst upland Georgian cotton sown on the same description of land gave 137¾ lbs. of clean cotton.

NOTE.—"These facts and figures show conclusively

that the Punjab cotton grower only requires good seed and suitable manure to produce four or five times as much clean cotton per acre as can be obtained at Cawnpore under the most favourable circumstances."—*J. F. Pogson.*

329. *Can you state how the cotton crops were manured at the Cawnpore farm?*

Yes. In his report, Mr. Bennett, Director of Agriculture for the North-Western Provinces, says that "on account of flooding by excessive rains, the experiments with manures were complete failures, because the action of the manures was disturbed by the abnormal rains of August and September (1881). They were washed from one field and deposited in another lying below it, and injured by excessive wet. The manurial experiments are therefore considered as valueless." (*Vide* Report.)

NOTE.—"Manures, except when applied as a top dressing, are dug into the soil and mixed with it. When applied to plants the same rule is observed. At Cawnpore unnamed descriptions of manure seem to have been spread over the surface of the soil as an experiment, and the unexpected removal thereof from one field to another below it, by the action of water, tends to show that the manure was of a light description; but how it came to be spoiled by being wetted is strange, for manure when dug into the soil requires to be wetted for its manurial value to be developed. Perhaps Mr. Bennett in some future report will afford us some further information as to the nature of his manurial experiment, so unfortunately brought to an end by neglect of surface drainage." —*J. F. Pogson.*

330. *Can you state what substances are present in cotton seed?*

Yes; 100 lbs. of cotton seed ash are said to contain, of potash, 34·75 lbs.; of soda, 1·10 lbs.; of lime, 6·0 lbs.; of magnesia, 13·73 lbs.; of oxide of iron, 0·55 lbs.; of phosphoric acid, 35·85 lbs.; of chlorine, 0·67 lbs.; of carbonic acid, 3·59 lbs.; and of sulphuric acid, 3·76 lbs.

NOTE.—"The above analysis is taken from the *Indian Agriculturist* for July, 1882. The name of the chemist is not given."—*J. F. Pogson.*

331. *Is cotton seed of any economic value?*

Yes. In America immense quantities thereof are pressed for their contained oil, which, being edible, is largely imported to Europe. The cotton seed cake is used as cattle food. In India, cotton seed is largely used as cattle food.

332. *What is the proper manure for cotton crops?*

Such crops need mineral manure, which should contain most of the substances mentioned in the analysis. A reference to the chapter devoted to manures will afford ample information on this subject.

333. *Can you state how the* Gossypium arboreum, *or tree cotton, also called the sacred cotton of the Deccan, should be cultivated?*

Yes. This almost unknown, but most valuable, variety of cotton is raised from seed, and the seedlings when twelve or more inches in height are to be transplanted, each plant being put down in well-manured soil; mineral manure being liberally used, at least two pounds being allowed to each plant. The seed bed is also to be manured, as ordered for common cotton.

As regards planting, one full-grown tree to one square

rod of land, or 160 trees to the acre, would be the proper number to put down. This cotton tree will soon attain the height of twenty feet, and as the lateral growth of the branches is a desideratum, ample growing room is necessary.

In favourable situations the *Gossypium arboreum* flowers and bears cotton pods within twelve months, and then continues to do so for over twenty years. In the Deccan the tree is said to yield a succession of cotton pods for nine months out of twelve, thus showing its superior value over the annual cotton plant. The cotton wool has been valued in London from from 6d. to 6½d. per lb. The plantation once formed becomes a remunerative property whose money value increases annually. In those parts of India where the strawberry will grow and ripen fruit, the ground under and between the trees may be put under strawberries; and where this may not be done, the ground-nut should be cultivated, and in like manner, when obtainable, *Chufas* should be similarly cultivated.

NOTE.—"With the large amount of convict labour always available, waste land could gradually be planted with *Gossypium arboreum*, and the plantations when formed could, and should, be leased to European and native capitalists. A square mile of tree cotton plantation would be a paying speculation, and as outdoor labour for convicts has been ordered by His Excellency the Viceroy, the suggestion, if viewed as a productive public work, is deserving of attention."—*J. F. Pogson.*

CHAPTER XXIII.

TOBACCO CROPS.

Introductory observations.—The tobacco plant, and components of green tobacco leaves.—The plant, how cultivated by the zemindar.—Improved method of cultivation suggested.—The leaves, how cured and prepared for the market.—Snuff, how made.—The yield of tobacco per acre.

INTRODUCTORY OBSERVATIONS.

OF all cultivated narcotics, tobacco is in use over the largest area, and among the greatest number of people. In 1492 Columbus found the chiefs of Cuba smoking cigars, thus affording proof that the cultivation of the tobacco plant, the proper curing of its leaves, the manufacture of cigars, and pipe tobacco, was well known to the Cubans; and when Cortes penetrated to Mexico he found tobacco in common use. The conclusion may therefore be drawn that the tobacco plant has been grown and used in America from time immemorial, and the fact of very ancient pottery tobacco pipes having been recently found, buried in the soil in the vicinity of prehistoric ruins, tends to prove the immense antiquity of tobacco culture in America. Under these circumstances it may be fairly admitted that all varieties of the cultivated tobacco plant have their home in America.

Tobacco, however, seems to have been known and used in India, when it was inhabited by a remarkably intellectual race of men, whose mother tongue was Sanscrit. In the inner Himalayas we have a wild variety of the tobacco plant, whose leaves and flowers very much resemble those of the Virginian tobacco. But this wild tobacco is a very hardy perennial, and withstands the severest frost, whilst all other known varieties of tobacco are annuals, and most suffer from the frost. It is quite possible that this wild tobacco was cultivated in the plains of India, and its leaves made into snuff and smoked in different ways. There are several Sanscrit words which indicate the various parts of the Indian water-pipe, known to Europeans as the "Hubblebubble," and tobacco was no doubt smoked in them. Hemp (*Cannabis sativa*) was also smoked. But as it had an especial form of bowl (*Chillum*) for holding the smoking preparation, the use of tobacco is indirectly proved. To take snuff would imply the presence of the snuff, and as the Sanscrit word *Nauslena* means to take snuff, the art of its preparation must have been known, and the cultivation of the tobacco plant understood and practised.

334. *Can you give a description of the tobacco plant, and its botanical, English, and Indian names?*

Yes. "It is a viscid herb; stem, three to six feet high, erect, round, hairy, branching at the top; leaves sessile (the lower ones decurrent), ovate, or lanceolate, acuminate, very large, pale green, with glandular short hairs; flowers in terminal panicles; calyx tubular, campanulate, five-cleft, hairy; corolla rose-coloured, funnel-shaped; throat inflated, ventricose; limb plaited, five-lobed, the segments acute; stamens five, included, of equal length; ovary two-celled; stigma capitate; capsule two-celled; seeds many, small, somewhat reniform, brown."—*Stephen and Church.*

NOTE.—The above is the botanical description of the *Nicotiana Tabacum*, Linn., or the Virginian Tobacco; and its Indian name is *Billaitee Tumaukhoo.*

335. *Can you state what substances are present in the green leaves of the tobacco plant?*

Yes. Ten thousand parts of tobacco leaves contain, according to the analysis of the chemists, Possett and Reimann, the following substances, viz.: 6 of the peculiar chemical principle nicotine; 1 of nicotianine; 287 of slightly bitter extractive; 174 of gum mixed with a little malic acid; 26·7 of a green resin; 26 of vegetable albumen; 104·8 of a substance analogous to gluten; 51 of malic acid; 12 of malate of ammonia; 4·8 of sulphate of potassa; 6·3 of chloride of potassium; 9·5 of potassa which had been combined with malic and nitric acids; 16·6 of phosphate of lime; 24·2 of lime, which had been combined with malic acid; 8·8 of silica; 496·9 of fibrous or ligneous matter; traces of starch; and 88·28 of water.

336. *How is the green tobacco leaf converted into marketable leaf tobacco?*

When the leaves attain full growth, the plants are cut, and hung up to dry during four or five weeks; they are then taken down out of the sheds in damp weather, for in dry they would be apt to crumble into pieces. The leaves are next stratified in heaps, covered up, and left to sweat for a week or two, according to their quality and the state of the season; during which time they must be examined frequently, opened up, and turned over, lest the leaves become too hot, take fire, or run into putrefactive fermentation. This process needs to be conducted by skilful and attentive operatives. An experienced negro can form a sufficiently accurate judgment of the temperature by thrusting his hand down into the heap. The tobacco thus prepared, or often without fermentation, is sent into the market.

NOTE.—"The above is the American plan of procedure, and accounts for the superior quality of the tobacco. The Indian plan is very different. The zemindar cuts his tobacco crop, and generally leaves the plants in the field, for the leaves to wither during the day, and to get bedewed at night. Next day the cut plants are removed and sun-dried, and when this operation is completed, the leaf tobacco is ready for use and sale.

"The zemindar has not the means to prepare leaf tobacco on the American plan; but if he was supplied with good seed, and grew the tobacco on properly manured land, and sold the crop as cut to European or native capitalists, who would undertake the proper preparation of the leaf, India would soon be able to produce large quantities of first-class leaf tobacco."—*J. F. Pogson.*

337. *How is the tobacco plant cultivated by the zemindars?*

In Upper India, before the rains commence, the tobacco seed is sown in seed beds, the seedlings when two or three inches in height are transplanted into growing beds, and when the rains have fairly set in the young tobacco plants are planted in the field prepared for their reception. In due time the crop is ready and cut, and the leaves prepared for sale.

338. *Can you suggest an improved method for the cultivation of tobacco?*

Yes. All municipalities should be supplied with the best kinds of country tobacco seed, the plants to be cultivated in the suburbs, and the land to be manured with night-soil, as explained in the chapter devoted to manures.

NOTE.—"When seed farms are established, all varieties of American tobacco, including the Cuban, Maryland, Kentucky, Virginia, and Orinoco, should be cultivated for the express purpose of producing an abundant supply of first-class acclimatized tobacco seed, to be supplied to municipalities for sowing on lands manured with night-soil."—*J. F. Pogson.*

339. *In what parts of India is the best country tobacco produced?*

Malwa produces a superior variety of opium, and the province includes the district of Bhilsah, which produces the best country tobacco, and it bears this name. Burmah also yields superior tobaccos, largely used for cigar manufacture. In the Bombay Presidency, Guzerat produces a fine variety of tobacco, to which the name of *Kaira* has been given by the growers.

NOTE.—"*Kaira* is evidently a corruption of Cairo. The tobaccos of Latakia (the ancient Laodicea), in Syria, and of Shiraz, in Persia, are famed for their superiority, and prized and priced accordingly, and their seed has only to be introduced into India to enable the agriculturists to produce these tobaccos in abundance. But until seed farms are called into existence this cannot be done."—*J. F. Pogson.*

340. *Is tobacco an exhausting crop?*

Yes. The tobacco plant takes more mineral matter out of the soil than does any other cultivated plant. It has been explained in a previous chapter that when vegetable substances are burned in the open air a mineral matter or ash remains. Professor Johnstone tells us, "The leaves of plants are especially rich in this incombustible ash, and those of the tobacco are among the richest in this respect among cultivated leaves. The dried tobacco leaf, when burned, yields from 11 to 28 per cent. of ash; or, on an average, every 4 lbs. of perfectly dry tobacco contains 1 lb. of mineral or incombustible matter." Further on we are told that "All the substances present in tobacco ash have been derived from the soil on which the plant was grown, and that they belong to the class of bodies which are at once most necessary to vegetation and least abundant in fertile soils. In proportion, therefore, to the weight of leaves gathered must have been the weight of these substances withdrawn from the soil. And as every ton (27 maunds) of perfectly dry leaves carries off 4 to 5 cwt. of this mineral matter—as much as is contained in 14 tons (378 maunds) of the grain of wheat—it will readily appear, even to those who are least familiar with agricultural operations, that the growing of tobacco must be a very exhaustive kind of cultivation."

NOTE.—"The analysis shows that 10,000 parts of tobacco leaves contain 20·6 parts of potassic salts, combined with malic and nitric acids. In addition 16·6 parts of phosphate of lime and 24·2 of carbonate of lime were present. Hence it follows that these substances should be largely present in tobacco manure; but nitre cannot be used, thanks to the Excise, and fossil phosphate of lime, though most abundant, is unattainable from want of ordinary cart-roads, to be followed by a railway, to link in the Sewallics with the Punjab railway at Saharunpore."—*J. F. Pogson.*

341. *How is snuff made?*

"The dried tobacco leaves are sprinkled with water, laid in heaps, and allowed to heat and ferment from one to six months. During this fermentation a chemical decomposition takes place in the leaves, and they give off at first nicotine and ammonia, and afterwards water and acetic acid. They are then reduced to powder, moistened with salt and water, and put into close boxes. Here they again heat and ferment. This gives them an agreeable ethereal odour and the well-known pungency of snuff."—*Johnstone.*

342. *How many pounds of tobacco will an acre of land produce?*

In Massachusetts, United States of America, 1,640 lbs. of tobacco is the usual yield per acre, and suitable land in India should produce the same.

CHAPTER XXIV.

MENSURATION.

Weights and measures connected with agriculture.—The land measures of England, how obtained from the standard square foot.—The Divine origin of these weights and measures demonstrated.—Circular measures of land.—Indian land measure.

THE English nation possesses standard weights and measures which were known to the inspired architect who planned and built the great pyramid of Ghizeh, in Egypt, whose carefully hewn stones and highly finished stone chambers have preserved the secret of these standards for—who can tell how many thousand years?

The prophet, priest, and great prince Noah was acquainted with these standards of mensuration, and used them when building the ark.

The children of Israel were not acquainted with them during their bondage in Egypt, and the Egyptians knew them not. But when the exodus took place under Moses, and the foot of Mount Sinai was reached, the Divine command went forth that a tabernacle should be constructed according to a measure of length called a cubit, and offerings of gold, silver, copper, and brass, when made, were to be weighed by a weight called the shekel.

The exact length of this cubit has long been a matter of dispute, and so has the weight of the shekel, which is another form of the ancient Arabic word *Miskal*, whose meaning is, the seventh part of an ounce. The correct word is *Sekel*, though it is not so written or pronounced.

That the length of the cubit had been lost is proved by the assertion of the Jewish historian Josephus, who states its length was equivalent to a standard then in use, and equalled 25·92 inches.

In 1638, Mr. John Greaves, Astronomy Professor Oxford, visited Egypt, and states that whilst in Cairo he took the measure of another cubit, which he met with in use in Egypt, and found it to be 1·824 English feet, or inches 21·888.

In 1685, Dr. Richard, the Bishop of Peterborough, wrote an "Essay towards the Recovery of the Jewish Measures and Weights," and in it he asserts that the ancient Hebrew cubit of the Mosaic dispensation measured exactly lineal inches 21·888.

The next cubit has been called the royal cubit of Memphis, and it measured inches 20·736; and we are told that a cubit of this proportion is commonly met with in the measures of the interior of the great Pyramid of Ghizeh.

The next and most remarkable measure found in this very ancient geographical landmark is the so-called Pyramid foot, which measures ·972 of the English foot, or 11·664 English inches.

The writer some years ago discovered the key which when applied to these ancient measures of length established the fact that the so-called cubit of 20·736 inches was no cubit at all, but a secret compound unit of mensuration, which when raised to a whole number gave a measure of length convertible into English feet of 12 inches, Hebrew or Tabernacle cubits of 18 inches, Hebrew or Tabernacle ells of 27 inches (*i.e.*, 1½ cubits), and mete yards of 36 inches each. But this was not all, for on one-tenth part of the measure of the sum so obtained being added thereto, the new sum gave a measure of length evenly divisible by inches 7·92, which is the length of each link of the standard land measuring chain of the Pyramid builder, next of ancient Israel, then of the race who inhabited prehistoric England, and now everywhere used in measuring land, and called Gunter's land chain.

The Pyramid foot of inches 11·664, under the operation of its key, gives similar results. But the most wonderful part of the discovery is that, the Altar of Incense, made to cubit scale by Moses, under Divine command, corresponds to Pyramid foot lineal measure, which was as unknown to Moses as it was to the Bishop of Peterborough; and it is clear that neither Pythagoras, Euclid, Diodorus Siculus, Josephus, nor Pliny fathomed the mysterious secret which it has fallen to the lot of the writer to penetrate and restore to mathematical science.

Before proceeding further it is necessary to explain that, in ages long gone by, compound and complex standards of lineal measure existed which, though perfectly understood by those in the secret, mystified and puzzled all others. We have examples of this in the cubit of Josephus and the royal cubit of Memphis, and by analyzing their length we arrive at their hidden and long-preserved secret.

Proof. The Egyptian or Peterborough cubit has the length of

Lineal English inches	21·888
Deduct length of Hebrew cubit inches	18·000
Sum in excess inches	3·888
Multiply this sum by	3
The product gives the length of the Pyramid foot, inches	11·664

In this example one-third of the length of the Pyramid foot has been united to the length of the true cubit of 18 inches, and has produced the puzzle now solved.

The cubit of Josephus measured

Lineal inches	25·92
Deduct length of true cubit inches ...	18·00
Sum in excess inches	7·92
Multiply this by	5
The product is inches	39·60

In this example the length of one link of the land chain has been added on to the length of the cubit measure, thus uniting the two standards; for we have only to deduct inches 3·6 from 39·60, the measure of the land yard, to obtain inches 36, the measure of the mete yard. If from this we deduct 9 inches we obtain the true ell of 27 inches, or a measure of three spans of 9 inches each. Then if from 9 inches of length we deduct inches 7·92, or one link, the remainder is inches 1·08; and if we multiply both sums by 25, the products are respectively inches 27, or true ell, and inches 198, or 16½ feet, or the length of one rod.

This interesting analysis has been given to show that the "span" contains the chain link, which is the one twenty-fifth part of the rod, and inches 1·08, which is the one twenty-fifth part of the ell. Hence any ancient Hebrew who was conversant with these secrets of mensuration (which Josephus was not) could at pleasure have given his nation their standards of mensuration, the units of which, though always present, were of as little use and value as if they had no existence.

We now come to the so-called

Cubit of Memphis, measuring inches	20·736
Multiply this sum by	125
The product is lineal inches ...	2,592·000

Here we have a measure just one hundred times the length of the cubit of Josephus, and the analysis teaches us that it contains 100 cubits of 18 inches each, and 100 chain links, the exact number present in the land measuring chain, of 66 feet, or 22 mete yards and 20 land yards. We thus obtain the most positive proof that the land chain, as a standard, existed before the great Pyramid was built, and we have now only to divide the whole number, inches 2,592, by suitable divisors to obtain quotients in exact accordance with England's system of lineal measure.

This Pyramid measure, inches 2,592, divided by 12 inches, yields 216 lineal feet, English measure; by 18, it gives true cubits 144; by 27, true ells 96; and by 36, mete yards 72.

Now these Pyramid measures, all emanating from the secret key measure of inches 20·736, are beyond cavil Pyramid measures, which existed perhaps before England was inhabited by man. But as they are all standard English measures, and were used by the children of Israel, who received them from Moses, who certainly did not invent them, it is reasonable to assume that he received his knowledge of them under Divine instruction.

It will be noticed that one peculiar standard, called the English ell, is not amongst the number. This ell gave its length to the Ark of the Tabernacle, which was 45 inches long.[1] To show that the true span was mathematically connected with the equilateral triangle, the square, and pentagon, we have only to remember that if one side of the triangle measured 9 inches, the three sides measured 27, and so produced the true ell; the four sides of the square gave 36 inches, or the mete yard; and the five sides of the pentagon produced the measure of 45 inches, which also exists within the Pyramid, but does not belong to the series of measures derived from the secret unit, 20·736.

It is now requisite to prove the connection between the Pyramid measures, of unknown antiquity, and the cubit and half cubit, or span measures, of the time of Moses; and to do so we have only to multiply inches 2,592 by inches 4·5 (the one-tenth part of 45 inches) to produce the cubical contents in inches of the Altar of Incense of the Tabernacle treated as a solid.

Proof lineal inches	2,592
Multiplied by inches lineal	4·5
Gives as product, lineal inches	11,664·0

[1] 100 cubits equal 1,800 English inches, which same divided by 45 gives as quotient 40 great ells of 45 inches, now called English ells.—*J. F. P.*

or one thousand Pyramid feet of English inches 11·664 each. This measure of length divided by 12 yields English feet 972, true cubits of 18 inches each 648, true ells of three spans each 432, and when divided by 36, mete yards 324.

The Altar of Incense was one cubit in length, one cubit in breadth, and two cubits in height. In other words, it was a double cube, and each of its sides measured 18 inches, or one cubit.

Proof calculation. Inches $18 \times 18 = 324$ square inches, which sum multiplied by the height, 36 inches, gives as product cubic inches 11,664; and a piece of string 972 English feet in length, if pegged down, would exactly equal the length of 11,664 one-inch cubes of wood, if they were to be placed alongside in a continuous line.

This calculation proves that the lineal measures communicated to Moses at Mount Sinai had, we know not when, been also communicated to Seth or one of his race, who built the Pyramid. We cannot tell why such great care was taken by the architect of the great Pyramid to hide all knowledge of the existence of these standards of mensuration, but that this part of his work was most scientifically and effectually done is undeniable.

The lineal measures of the Pyramid of ancient Israel, and of prehistoric as well as historic and modern England, have been placed before the reader, who, however does not know that the prime unit of all these measures is the primal cube, and that its six sides give the first measure of length. It divides the length of the foot into 100 parts which equal 12 inches, each of which is subdivided into 10 parts; hence 120 such parts exactly equal the length of 100 primal cubes, or one foot of length.

The primal cube measures the one-hundredth part of the foot of 12 inches squared and cubed, and it is the parent of the foot and all other measures square and cubical; and, in addition, a cube vessel of capacity, constructed according to primal cube scale, when filled with pure undistilled water at an ordinary temperature, will produce the water ounce (measure), weighing 437½ grains, and all its ancient and long-lost subdivisions, called in Scripture "the weights of the bag," and which were declared by King Solomon to be of Divine origin.

The six sides of the primal cube used as a lineal measure produce six-hundredths ($\frac{6}{100}$) of the foot, or, in inch measure, 0·72 inches. From this measure, or first standard, spring all the measures of length already detailed, and in addition it gives birth to mathematical mensuration in its various forms.

For example: Inches 0·72 × 10 = inches 7·2 or one-fifth part of the mete yard. Now from this apparently insignificant measure of length is developed the first and most ancient problem in mathematics and geometry, or land measure.

Demonstration: Inches 7·2 and primal cubes 60 equal each other and give us six equal parts of ten primal cubes each, or, in inch measure, six equal parts of inches 1·2 each.

If we call inches 7·2 the perpendicular of a right-angled triangle, the two other sides will measure inches 9·6, or primal cubes 80, and inches 12, or primal cubes 100. Hence it is demonstrated that the foot of 12 inches is the hypothenuse of a right-angled triangle, whose base is inches 9·6, and perpendicular inches 7·2.

If we square the perpendicular, and mark every tenth primal cube on each of the square sides, and then connect them by lines crossing each other at right angles, the square will be divided into 36 square aliquot parts, each side measuring inches 1·2. The square of the base similarly divided will yield 64 such parts, and the square of the hypothenuse 100 similar parts. Now 36 squares on being added to 64 squares produce 100 squares, which equals the measure of the square of the hypothenuse.

We thus obtain the first lesson in superficial square inch measure. The hypothenusal square contains 144 square inches, or one square foot, the basic square 92·16 square inches, and the square of the perpendicular 51·84 square inches, which being added together produce 144 square inches, or a second square foot, and we thus obtain a parallelogram whose opposite sides measure 24 inches and 12 inches respectively.

The next lesson to be learned from this problem is cubical mensuration, as connected with measures of capacity and weight.

A vessel or water-tight box whose internal dimensions measure 12 inches in every way will contain 1,728 cubic inches of pure undistilled water; and as the square foot was superficially divided into 100 parts, the cubed foot contains 1,000 such aliquot parts in the cube form.

The cubic foot of pure water weighs 1,000 ounces, and hence it follows that each water cube weighs one ounce. A vessel whose internal dimensions measure inches 1·2 every way will exactly contain one ounce of water in the cube form, and the weight of the water so impounded will be 437½ grains.

This cubed water ounce measure is exactly divisible into 1,000 parts, each having the dimensions of the primal cube, and a minute box of its dimensions will hold exactly seven-sixteenths of a grain of water.

To prove this, we convert 437½ grains into sixteenths, and divide by 7, when the quotient will be 1,000 primal cube measures.

Then 437½ × 16=7,000 sixteenths, which sum divided by 7 gives 1,000 primal cubes as the quotient; and we thus learn that the half-grain in this pure water ounce, which has so puzzled ancient and modern wise men, is essentially necessary to fill up the cube ounce measure, and this 1,000 times repeated equals 500 grains per cubic foot of water, and therefore a very sensible quantity.

From this water ounce is derived the shekel of the Tabernacle of the time of Moses, and its exact weight was grains $62\frac{8}{16}$ and its half (called *Bekah*, the soul ransom whenever the children of Israel were numbered) was a key weight, whence the troy tithe ounce of 450 grains and the troy ounce of 480 grains were derived.

Proofs. The half of grains 62·5 is grains 31·25. Then grains 31·25 × 15·36=480 grains, which is the weight of the troy ounce. Next 31·25 × 14·4=450 grains, which is the weight of the tithe ounce belonging to the troy system, and it was obtained by tithing 150 ounces troy, or pounds 12½ troy.

Proof. Grains 480 × 150=72,000 grains, whose tenth part or tithe is grains 7,200, which sum divided by 16 gives 450 grains as the weight of its ounce, and 7,200 grains as the weight of this pound.

The shekel was further subdivided into 20 gerahs, weighing 3⅛ grains each.

In India the double troy tithe pound, weighing 14,400 grains, is called a seer, and its thirty-second part, or grains 450, is officially denominated the "postal ounce." The double ounce is called the chittack, and the standard maund contains 40 seers. The weight of the Indian rupee is 180 grains, or one tola, and 80 of them officially represent one seer in weight.

The zemindars of the Bengal Presidency, and all vendors of food supplies, purchase and sell by these weights.

We have thus step by step brought down the sacred measures and weights which have existed on this Earth at a period of antiquity anterior to the building of the great Pyramid (as proved by the stone ball, weighing grains 8,325, found by Mr. Piazzi Smyth in a closed secret chamber within the Pyramid, which ball was first of all placed within it, and then the chamber was closed by the course of stone masonry forming its roof. This ball is now in the British Museum, and its weight under analysis yields all the key weights belonging to what is now called the avoirdupois, the troy, the tithe, and the tola system of weights) to the year of grace 1882, and trust we have for ever exploded the fiction that the length of three barleycorns produced the English inch which in its turn supplied all the other standards of mensuration.

All requisite data being at the critical as well as the practical reader's disposal, we may now proceed with the land measures of England, which were in use as such when the children of Israel, under the leadership of

Joshua, took possession of their allotments of land in Palestine.

These land measures are also of Divine origin, and tend to prove that one and all represent the areas of circles thrown into the form of squares and parallelograms for the convenience of the uneducated agriculturists of Israel.

The series begins with—first, the lineal land yard of five chain links of inches 7·92 each, or inches 39·6, which five times repeated or measured produced the land rod, measuring 25 chain links in length, or 198 inches or 16½ feet; and this quantity squared gave 25 square land yards which equalled square feet 272·25, and square inches 39,204.

The most ancient grain measure to found in England is the "Winchester corn gallon," which contains 272·25 cubic inches. In this standard we have a model of the land rod on the scale of one inch to the foot, inasmuch as 16½ inches squared will produce 272·25 squared inches. Hence it follows that a board one inch thick and 16½ inches square will represent 272·25 cubic inches; and with this data to hand we may easily construct a vessel or box which will hold the gallon, half-gallon, and quarter-gallon (quart) of wheat.

The true water gallon contains cubic inches 276·48, the corn gallon cubic inches 272·25. It is quite possible that a square foot of soil sown with a single grain of seed wheat may produce grain sufficient to fill a cubic inch measure, and if so, the very practical nature of this ancient grain measure of the children of Israel is demonstrated. But how this, and all the other weights and measures of the ancient pure-blooded Hebrews—which were in daily use in Palestine when Isaiah lived, wrote, and prophesied B.C. 723, and King Hezekiah ruled in Jerusalem—found their way to England, may at some future date be made known in a work exclusively devoted to the subject, and now extant in manuscript awaiting publication.

THE LAND MEASURES OF ISRAEL AND ENGLAND.

The dimensions have been given of a square plot of land called a square rod, which represents the one hundred and sixtieth part of a standard block of land called the acre, and which has *ab initio* been subdivided into ten equal square parts called square chains, and this division still holds good. The acre was further divided into four parts or quarters, called roods.

To show that all these square and oblong blocks of land were derived from the areas of their respective circles, the formula for laying out the rod circle, the squared chain circle, the rood circle, the acre circle, and the acre in annular or ring form, are given in succession, and these in ancient times were no doubt in use as part of the land measure system of Israel, as well as for the construction of works of irrigation, such as wells, reservoirs, cisterns, and large circular ponds.

Rule: Multiply the diameter by the true or sacred ratio, inches 3·13324, to obtain the circumference; and half the circumference multiplied by half the diameter will give the area of the circle in square inches and square inch fractions.

No. 1.—The Rod Circle.

Diam. Inches	223·71667002
Mult. by Ratio In.	3·13324
	2)700·9580191734648 = Circumf.
½ Circumf.	350·4790095867324
½ Diam.	111·85833501
Area of Circle	39203·99846832571445o421324
Minus	0·01
Sum needed	39204o·0 Squared inches.

The Rod's length is Lineal Inches	198
Mult. by L. Inches	198
Product, Square Inches	39204

No. 2.—The Squared Chain Circle.

Diamt. Inches	894·86668
Mult. by Ratio L. Inches ...	3·13324
	2)2803·8320764432 = Circumf.
½ Circumf.	1401·9160382216
Mult. by ½ Diam. ...	447·43334
Area of Circle	627263·9753810581481 44
Minus	0·03
Sum needed	627264·0 Squared Inches.

The Land Chain Measures, Inches	792
Which Mult. by Lineal Inches	792
Gives Product, Square Inches	627264

Rediscovered and Calculated by

J. Fred Pogson.

Simla, 21*st Sept.*, 1878.

No. 3.—The Rood or, ¼-Acre Circle.

Diam. Inches ... 1414·908483

Mult. by In. ... ... 3·13324, the Sacred Ratio.

2)4433·24785527492=Circumf.

½ Circumf. ... ... 2216·62392763746

½ Diam. ... ... 707·4542415

Area ... ... 1568159·9994175101512865590

Minus ... ... 0·0006, lost by Fractions the $\frac{8}{10000}$ of a Square Inch.

Sum needed, Sq. In. 1568160·0000=One Rood.

The rood divided diagonally into two equal parts will give an isosceles triangle, whose base will measure 44 mete yards, and perpendicular 55 mete yards. The rood is a parallelogram, whose sides measure 55 mete yards and 22 mete yards respectively. Hence the area of the isosceles triangle and the rood circle equal each other.

No. 4.—The Acre Circle.

Diam. Inches ... ... 2829·816966

Mult. by Ratio In. ... ... 3·13324

2)8866·49571054984=Circumf.

½ Circumf. ... ... 4433·24785527492

½ Diam. ... ... 1414·908483

Area of Circle ... ... 6272639·99767004060514636

Minus ... ... 0·01, lost by Fractions.

Acre equals ... ... 6272640·00 Square Inches.

No. 5.—The Acre in the Form of a Square.

Side of Square Lineal Inches ... ... 2504·523905

Multiplied by L. Inches ... ... 2504·523905

Product, Square Inches ... ... 6272639·990716449025

Minus ... ... 0·01

Area of Square Acre, Sq. In. ... ...6272640·00

The standard acre is a block of land which measures 220 mete yards in length, by 22 mete yards in width. It contains 4,840 squared mete yards, and 4,000 squared land yards. It also contains 43,560 square feet, of 144 square inches each, which equal 6,272,640 square inches. The squared land chain contains 484 square mete yards, and constitutes one-tenth part of the acre.

No. 6.—The Three-Acre Circle.

Diam. Inches	4901·386752
Mult. by	3·13324
	2)15357·22102683648=Circumf.
½ Circumf.	7678·61051341824
½ Diam.	2450·693376
Area ... ... Sq. In.	18817919·922118039885578 24
Minus	0·08
3 Acres equal	18,817,920 Square Inches.

No. 7.—Four-Acre Circle.

N.B.—Each Quadrant equals one Acre.

Diam. Inches	5659·633932
Mult. by L. In.	3·13324
	2)17732·99142109968=Circumf.
½ Circumf.	8866·49571054984
¼ Diam.	2829·816966
Area ... ... Sq. In.	25090559·99068016242058544
Minus	0·01
4 Acres=Sq. In.	25090560·00 ÷ 4=6272640·0

No. 8.—The Acre in the Ring or Annular Form.

From 4-Acre Circle Diam. In.	5659·633932
Deduct 3-Acre Circle Diam. In.	4901·386752
The difference halved gives the	2)758·247180 ÷ 2
Width of the Ring, L. In.	379·123590

No. 9.—THE SIDE OF THE FOUR ACRE SQUARE

Measures Lineal In. 5009·047810 multiplied by Inches 5009·047810, giving Square Inches 25090559·9628, &c., or four-hundredths of a Square Inch short of 4 Acres in the form of a Square.

The square inch is easily divided into one hundred parts, and as in the area of the acre circle less than one such part is wanting to complete the last square inch, the remarkable accuracy of the ancient system of mensuration must be admitted as practically converting the area of a square into the area of a circle, and *vice versâ*. This was due to the ancients being acquainted with the formula for correctly calculating the circumference of the circle from the diameter. This problem has been called the "rectification of the circle," and an example is given beneath to prove the assertion.

One degree of the circle's circumference equals nine inches. Required the diameter, and circumference by calculation.

Then degrees $360 \times 9 = 3,240$ inches lineal.

Diam. Inches	1034·073355
Multiplied by Ratio Inches	3·13324
Gives as the Circumference Inches	3239·99999882020
Add lost by Fractions	0·00000117980
Circumference by Measure	3240·00000000000

By this calculation the sum short is less than two-millionths of a lineal inch.

The same diameter, 1034·073355, multiplied by ratio inches 3·1416, gives as the circumference inches 3248·644852068o, or inches 8·644 in excess of the circumference by measure, inches 3,240.

NOTE.—"The circumference of this circle, lineal inches 3,240, is a standard measure, and contains all the ancient measures of length, which though lost for ages to the Hebrew nation, have for ages and up to this day been preserved in England. If lineal inches 3,240 be divided by 45, the quotient is 72 great ells, now called English ells. By 27 we obtain 120 Hebrew ells, now called Flemish ells. By 18 we have 180 Hebrew cubits which equal 90 mete yards, and divided by 12 we obtain 270 feet English.

"If to inches 3,240 we add its one-tenth part, inches 324, the product is inches 3,564, which divided by 792 gives as quotient 4 chains and a half. Thus we own, as national standards, measures of length which existed before the Great Pyramid was built."—*J. F. Pogson.*

The antiquity of all the various measures under review having been explained as well as established, the table given has been drawn up for use and reference.

Table of the most ancient measures of length, now called "Long Measure."

Primal Cubes.		Inches.	
1.	The Unit of all Mensuration ...	0·012	
6.	First Measure of length	0·072	
60.	First Perpendicular, height ...	7·2	Is ⅕ of the Mete Yard
66.	First Land Measure, length ...	7·92	Is ⅕ of the Land Yard
75.	First Span	9·00	Is ⅕ of the Great Ell
80.	First Base of R.A. Triangle ...	9·6	Is ⅕ of 4 Feet
100.	First Hypothenuse of Triangle ...	12·0	Is called the Foot
150.	First Double Span Measure ...	18·0	The Hebrew Cubit
225.	First Ell or Triple Span	27·0	The Cubit and a half
300.	First Mete Yard	36·0	The Double Cubit
330.	First Land Yard	39·6	The Yard of 5 Chain Links
375.	First Great Ell of 9 Spans... ...	45·0	Is now called the English Ell
600.	The Double Mete Yard	72·0	The Fathom
660.	The Double Land Yard	79·2	The Toise
825.	The Half Land Rod	99·0	The Reed
1650.	The Land Rod	198·0	The Rod of 16½ feet
6600.	The 4-Rod Land Chain	792·0	Is now called Gunter's Chain

The furlong of 220 and the mile of 1,760 mete yards are chain measures of length, and these standards squared produce the square or superficial measure belonging to each. The square mile contains 640 acres.

The mile is also the hypothenuse of a right-angled triangle, whose perpendicular measures 1,056 mete yards in length, and base 1,408 mete yards in length, thus proving that the mile is a strictly mathematical and geographical measure of length.

Table of the most ancient Cube Measures of capacity which yield the Water-weight system of Weights and Measures, commonly called the Avoirdupois system.

Primal Cubes 10, equal in length Lineal Inches	1·2
Primal Cubes 10, Squared, give 100 Square Primal Cubes, which equal Squared Inches...	1·44
Primal Cubes 100, Cubed, yield 1,000 Primal Cubes, which equal Cubic Inches	1·728

A vessel of cube form made of metal and measuring internally 1·728 cubic inches will hold exactly 437½ grains of pure undistilled water, or one water ounce in the cube form.

If the four sides of this box be made of metal having the thickness of half a primal cube, or the two-hundredth part of the lineal foot, each side will measure lineal inches 1·32. The bottom or base being made of metal, one primal cube, or one-hundredth of a foot in thickness, will produce a cube measure whose external height will be that of eleven primal cubes, or lineal inches 1·32, and this sum squared will yield the area of its base or squared inches 1·7424. This product is the square unit of the land measure system, whilst the inside measure of the bottom or base of the vessel, squared inches 1·44, belongs to the square foot system; and as both are present in the water-ounce measure, the connection between the weight, or one ounce, and the vessel impounding the water is established.

From this first cubed water ounce is derived the half water pound, which is also a cube of eight cube water ounces, and its sides measure respectively lineal inches 2·4, whilst the diagonal of its square and cube measures lineal inches 3·39.

A box whose internal dimensions equal inches 2·4 in height, breadth, and width will hold exactly eight ounces of water. The half-pound water measure or vessel contains 13·824 cubic inches, and the pound weight of water, in any form, equals 27·648 cubic inches in bulk.

With the half-pound to hand it is easy to construct all the weights which emanate therefrom, namely, the pound of 7,000 grs., the clove of 7 lbs., the stone of 14 lbs., the quarter of 28 lbs., the hundredweight of 112 lbs., and the ton of 2,240 lbs., or 20 cwt.

The water ounce is the parent of all these measures of weight, and in the time of King Solomon, the great sovereign of the children of Israel, it was subdivided in a particular manner, the resulting series of small weights being called the "weights of the bag."

Of these, and the other weights belonging to the system, King Solomon stated: "A just weight and balance are the Lord's; all the weights of the bag are His work" (Prov. xvi. 11).

These weights have been lost since the destruction by fire of the Temple at Jerusalem, B.C. 587, in the nineteenth year of the reign of Nebuchadnezzar king of

Assyria, and their recovery and restoration by the writer is due to the discovery of the primal cube.

No.	THE WEIGHTS OF THE BAG.	Grains.
1.	The weight of the Primal Cube	$0\cdot\frac{1}{16}$
2.	The weight of the Key Cube $\frac{56}{16}$	$3\cdot\frac{8}{16}$
3.	The weight of the Double Key Cube of 16 Primal Cubes	$7\cdot$
4.	The weight of $\frac{1}{20}$ of the Ounce	$21\cdot\frac{14}{16}$
5.	The weight of $\frac{1}{10}$ of the Ounce	$43\cdot\frac{12}{16}$
6.	The weight of $\frac{1}{8}$ of the Ounce	$54\cdot\frac{11}{16}$
7.	The weight of $\frac{4}{20}$ of the Ounce	$87\cdot\frac{8}{16}$
	Total Grains	$218\cdot\frac{12}{16}$

NOTE.—"These small weights in the aggregate weigh half the weight of the water ounce; and weight No. 4 added to weight No. 7 produces the quarter-ounce weight of $109\frac{6}{16}$. Hence the seven weights of the bag contain all the fractions of the ounce, and as from any one of them the water ounce could be found, their Divine origin, as well as of all the standards of mensuration, is undeniably established.

"In India the land is measured by the Beegah,[1] which is a local and variable quantity. In some parts of the country three beegahs equal one acre, and in others five. The agriculturist should know how to calculate the yield of his land, and the information given in this concluding chapter of our work is practical, comprehensive, and exhaustive."

[1] The Beegah is subdivided into 20 parts called Biswas.—*J. F. P.*

J. F. POGSON.

DEHRA DOON,
17th November, 1882.

APPENDIX.

WHEAT.

THE introduction into India of superior varieties of imported wheat was opposed by the Agricultural Department of the North-Western Provinces because such wheat does not answer at Cawnpore !

The wealth of an agricultural country in a great measure depends on the superior quality and quantity of the cereals which it can produce for exportation, and of these wheat is the most important, as it always commands the highest money value ; superior barley comes next, maize and *Doorra* or *Jowar* follow, and rice terminates the list.

The successful cultivation of superior varieties of imported wheat is intimately connected with the soil and climate of the locality.

In Upper India, a good wheat soil and an unsuitable climate cannot be expected (except by tiros in agriculture) to produce a good wheat crop, and a failure may always be safely predicted if the seed wheat has been obtained from Palestine, Russia, Spain, France, and Great Britain.

Wheat is in India always a cold weather crop, and those parts of India which have the longest winter, as a rule, produce the best indigenous wheat crops, and are the best fitted for the cultivation and acclimatization of those European varieties which command the highest prices in the London market. It is for this (climatic) reason that the white and red wheats of the Simla district are so greatly superior to the grain of wheat of the plains.

The acclimatization of such European varieties cannot be carried out in the plains, except perhaps in the northern districts of the Punjab, beginning at Hoshiarpore and ending at Rawul Pindee.

The Punjab hill stations north of the river Jumna, and the adjacent wheat lands *en route*, are all more or less adapted for the successful acclimatization of the first sowing of all European varieties of wheat. But on this, or eastern side of the Sutledge river, only two localities in British territory are available for this purpose, the first being at and about Subathoo, and the second in the Simla sub-district of Kotgurh. The formation of Government seed farms (as contradistinguished from experimental farms, such as those of Cawnpore and Nagpore) at both places, and especially at the latter, is a desideratum, inasmuch as all altitudes, rising from 6,400 feet above the level of the sea to 10,000 feet, and descending from 6,400 feet to 2,800 feet are there to be met with; and as the grand slopes are not precipitous (like the Mussoorie hills) wheat, barley, wheat-barley, and maize are largely cultivated. In fact no place on this (eastern) side of the Sutledge river, from its most favourable aspect and latitude, is better fitted for the formation of a seed farm.

In Kulu and Kangra similar seed farms should be formed, and these, with the Hoshiarpore, Jhelum, and Rawul Pindee seed farms, would enable the Punjab to become what it is destined to be—the granary of Great Britain.

To show the necessity which exists for the formation of seed farms for the purpose indicated, I will quote Mr. Fuller, who, not being able to account for his failures in imported seed wheat cultivation, consoles himself by singing the praises of Cawnpore bazar wheat.

In his report, Mr. Fuller, Assistant-Director of Agriculture North-Western Provinces, late in charge of the Cawnpore Experimental Farm, states: "In face of the excellence of indigenous wheats and barleys, the attempt to introduce English varieties possesses little practical interest. It is worth noting, however, that this is the first season in which any produce whatever has been gained from English wheat seed, the trials of the two years preceding having resulted in complete failure. It will be interesting to see whether the acclimatized produce will give a better return next year.

"The varieties cultivated were, first, the English rough chaff, which gave a yield of two bushels per acre; and second, the English red-bearded, which gave 4¾ bushels per acre." It appears that imported seed wheat "does not seem to become acclimatized

GOLDEN GRAIN, OR PALESTINE MAMMOTH WHEAT.

WHITE RUSSIAN SPRING WHEAT.

in less than three years, so that the first two years' experience is generally disheartening." (*Vide* Report.)

Now, had these wheats been sown and grown at Subathoo and Kotgurh, they would at once have matured acclimatized seed. But when it is known that the hot weather begins at Cawnpore in March, and that English wheat does not usually ripen till May—June, it follows that a hot February would act injuriously, and March destroy the standing crop without its seeding. In the Punjab the wheat of the country ripens seed in April and May, and in Kotgurh in May and June, according to altitude; hence English wheat of all kinds would succeed at once in the Himalayas, and the acclimatized seed in the Punjab, but never in Cawnpore, with its short winter followed by its well-known heat.

The Polish, or "forty days to sixty day's wheat" (*Triticum Polonicum*), would no doubt suit the climate and latitude of Cawnpore, and as this wheat has a standing in the markets of Europe and London, its successful production would compensate the Experimental Farm officials for past disappointment and failures.

The "Golden Grain, or Palestine Mammoth Wheat," of which an illustration is given on page 269, is thus described by Mr. James J. H. Gregory: "The grains are really enormous, being three times as large as those of common wheats, and therefore worthy of being put on the market as a curiosity were it nothing more; but the grains are not only mammoth in size, but the wheat crops wonderful, having yielded last season thirty-two bushels on half an acre of land. The first grain in this country (America) was brought from Palestine four years ago. It is a very hardy and most productive variety, with long stiff straw standing well; stools heavily, growing from 25 to 70 stalks from a single grain; the ears are of good length, closely filled with full, plump grains with but little chaff."

This wheat has been introduced into the Punjab by the writer, and it is growing in his garden. The supply received was small, under 400 grains, of which 150 were sent to Mr. Edgar Spooner, Superintendent, Agri-Horticultural Society's Gardens, Lahore.

The second illustration is that of the "White Russian Spring Wheat." Of it we are told: "No wheat ever tried in this country (America) has received a more unanimous commendation from those who tried it. Has such long, strong, healthy yellow straw, standing straight several days after ripening, and bearing large, long white chaff heads, well filled with plump kernels, weighing oftentimes from 60 to 62 lbs. to the measured bushel. Its yield is from 45 to 50 bushels per acre." This variety has not yet been introduced into India. The other varieties demanding notice and introduction are, first, the French Imperial Spring Wheat; second, Champlain Wheat; and third, Defiance Wheat.

If India is to compete with America and Russia in the production of superior wheats, that result can only be obtained by sowing similar varieties of seed. The best wheat now produced in India will never equal that of Palestine, and the best wheat produced in Upper India and the Himalayas will not become better by being cultivated at Cawnpore. That it may degenerate is more than probable.

WHEAT-BARLEY.

THE cultivation of this valuable variety of barley was opposed by Mr. Bennett, Director of Agriculture, North-Western Provinces, because the Cawnpore Farm officials mismanaged its culture.

The wheat-barley condemned by Bennett has been successfully grown at Dehra Doon and in the Agri-Horticultural Society's Gardens at Lahore, and also at Goruckpore and Bijnor, in the North-Western Provinces.

It is one of the peculiarities of Indian officials (with rare exceptions) to oppose, tooth and nail, improvements of all kinds, if initiated by an outsider. The contest between the writer and Mr. Bennett is an example of this nature, and but for the Press he would have carried the day.

"THE WHEAT-BARLEY OF THIBET.

"*To the Editor of the 'Civil and Military Gazette,' Lahore.*

"SIR,—Mr. Bennett, the Director of Agriculture, North-Western Provinces, enjoys the privilege of having his agricultura successes and failures published in the *Government Gazette* of the North-Western Provinces, and he has recently been pleased to make a perfectly uncalled-for personal attack on me, for having brought the very valuable wheat-barley of Thibet to the notice of the Secretary to the Government of India, in the Agricultural Department, the result being that about two maunds of seed wheat-barley (white and dull green varieties), value about five shillings, were sent by me from Kotgurh to Simla, and from Simla, under Mr. Buck's orders, to the Superintendent of the Cawnpore Farm, for cultivation. An official letter in my possession intimated that the seed had arrived too late (no fault of mine) for sowing. It appears, however, that the wheat-barley was sown, grown, and harvested, and the culture having somehow been misdirected, Mr. Bennett turns round and attacks me in his report, in place of the Cawnpore Farm officials. In his wrath, Mr. Bennett gives vent to his feelings in the following language, viz.: 'The naked barley, known as *rasuli*, or *paigham bari*, the excellence of which has been highly vaunted by Captain Pogson, gave an outturn of only 10 maunds (13·3 bushels) to the acre, and is,' as Mr. Bennett describes it, '"an impostor."' Mr. Bennett, in six lines of print, has made no less than three ridiculous blunders. First. I only stated the bare fact of the Kotgurh zemindars holding the wheat-barley in high estimation, on account of the fine meal which it yielded, and the good bread *(chuppatties)* which such meal made. The absence of husk gave a larger yield of flour, hence a maund of wheat-barley contained considerably more breadstuff than the like quantity of common husk barley. A simple and correct statement of facts cannot under any circumstances be regarded as a vaunt, or vain boast, as ascribed to me by Mr. Bennett. Second. Mr. Bennett states that the wheat-barley only gave a yield of 10 maunds, or 13·3 bushels. Now, as 10 maunds equal 820 lbs., and as a bushel of the best English barley weighs 46 (forty-six) lbs., the yield in bushels is 17¾ and 3½ lbs. Can it be possible that the Director of Agriculture, North-Western Provinces, has not yet learned that a bushel of wheat weighs 60 lbs., one of oats 32 lbs., one of bran 27 lbs., and one of barley 46 lbs.? 820 lbs. divided by 60 lbs. will give bushels 13⅔, and not 13·3, as erroneously stated by Mr. Bennett. Third. Mr. Bennett officially proclaims the wheat-barley to be an impostor, and in reply I adduce crushing facts to demolish both the assertion and the assumption. In October, 1881, Mr. P. Sri Lall, Secretary, Bijnor Agricultural Society, was supplied with two maunds of wheat-barley (varieties white and dull green), the seed being the same as that which officially found its way to Cawnpore. On the 21st May, 1882, Mr. Sri Lall wrote to me as follows: 'Both varieties of the wheat-barley, namely, the dark-coloured and the indigenous wheat-barley, were tried, and their outturns are 22 maunds, 15½ seers, and 4 maunds and 6 seers per acre, respectively. It will thus be seen that the dark-coloured barley has given highly satisfactory returns; the miserable yield of the other variety is mainly due to the fact that it was cultivated on sandy soil (*bhor* land), no other plot being available when the seed arrived. More careful experiments would, however, be made this year.'

"In the Proceedings of the Agri-Horticultural Society of India for April, 1882, appears the communication given below, being sent by Mr. W. C. Peppe, of Birdpore, Goruckpore: 'The wheat-barley promises to be a great success. It was the first ripe, and first cut, of any *rabi* crop about here. The standing crop was a magnificent sight. The produce from four-fifths of an acre manured with bone-dust was 1,156 lbs.; straw 1,640 lbs. The produce of two-fifths of an acre manured with ordinary farmyard manure was 255 lbs. grain, and straw 410 lbs.' Mr. Nicholson, of Mirzapore, originally supplied the seed wheat-barley to Mr. Peppe. The first handful came from Nepal, and was brought by a pilgrim visiting Benares. Mr. Peppe obtained 17 maunds 23¾ seers of wheat-barley per acre, and considered it a great success. Mr. Sri Lall harvested 22 maunds and 15½ seers, and both have supplied valuable information as to why the wheat-barley did not answer when sown on unsuitable soil and improperly manured. His Honour the Lieutenant-Governor of the Punjab (Sir H. Egerton) caused the Lahore Agricultural Society to be supplied with 18 seers of the Kotgurh wheat-barley for experimental sowing, and ordered 22 seers to be sown on the well-manured land belonging to the grounds of Government House, Lahore. But the results are unknown to me. I may now mention that the wheat-barley (three varieties, white, dull green, and brown; I had samples of each, sent to me from Poo, in Thibet, by my good friend the Rev. E. Pagel,[1] missionary in that part of Thibet) is very extensively cultivated in Thibet, the climate being too cold for wheat, whilst the common, or husk-barley, is only cultivated on a small scale as cattle food. In China wheat-barley is also much cultivated, yet this valuable cereal, which is the national food of the Thibetans, and has been so from remote antiquity, is absolutely condemned by Mr. Bennett, and I am held up to public ridicule for having brought it (wheat-barley) to the notice of the Government of India, in the Agricultural Department. My object in so doing was to improve the barley crops of the plains, and to induce the zemindars to give up cultivating the common husk barley.

"J. FRED POGSON.

"Mussoorie, 28th Sept., 1882."

[1] Jan. 1883. The death of Mr. Pagel, and also of his wife, after twenty years' residence, has just been announced in the *Simla Argus*.—*J. F. P.*

The above was published on the 10th October, 1882, and a few days after the writer received a copy of "Report on the Workiug and Condition of the Punjab Agri-Horticultural Society's Garden Lahore, for the year 1881–82," and the extract given beneath is taken therefrom.

"The seed of the variety named 'Huskless, or wheat-barley,' was received from the Private Secretary to Sir Robert Egerton for experimental cultivation in the garden with a view to its acclimatization in the plains.

"The variety was brought to notice by Captain Pogson, who found it at Kotgurh, in the Simla District. The yield of this variety, both in grain and straw, as will be seen from the table given, is larger than the yield of either of the other varieties tried. The yield from the heaviest manured plot being 24 maunds grain and 48 maunds straw per acre. The plot manured at the rate of 200 maunds per acre produced at the rate of 27 maunds 24 seers of grain and 39 maunds of straw per acre. Of oats, the variety named 'Red rust proof' gave the largest outturn, as it also did in 1881. This variety yielded at the rate of 13 maunds 32 seers of grain and 78 maunds of straw per acre; the yield of oats 'Board of Trade' being at the rate of 9 maunds grain and 84 maunds straw. The plants in this plot averaged upwards of 6 feet in height. The grain of the 'Huskless, or wheat-barley,' harvested is of an inferior description to the seed sown; it is small and discoloured, having the appearance of being over-ripe. Further experiments will be made with this variety during the current year. Earlier sowings will be made to enable the crop to mature and be cut before the heat in the plains is so great."

The manure so largely used was night-soil, and as the unmanured plot gave 19 maunds and 32 seers of grain and 34½ maunds of straw, we have a practical proof of the unfitness of night-soil as a manure for wheat-barley. The grain has suffered in consequence. Had proper manure been used 39½ maunds should have been produced, or double the quantity of grain obtained from the unmanured plot. Mineral manuring has been suggested to the superintendent, and will be adopted.

About ¼ lb. of Kotgurh wheat-barley was given to Mr. Guthrie of Dehra Doon, in November, 1881, and sown by him between the 20th and 24th of that month. The seed germinated quickly, and though sown fully a month later than the common barley growing in an adjacent field, the crop ripened seed one week before the common barley. The yield was abundant, and the grain was fully equal to the seed sown.

To ascertain the food value of this grain, a tin canister was filled with common husk-barley purchased in the Mussoorie bazar, and weighed; the weight was exactly 6½ oz. The same tin was then filled with the wheat-barley grown by Mr. Guthrie, and it weighed 7½ oz. and 60 grs. If from the husk-barley we deduct from 20 to 25 per cent. of its weight for the husk, the remainder is the food value of the grain. At 25 per cent. we have 4⅞ oz. of food (barley-meal) and 1⅝ oz. of husk. The difference in favour of the wheat-barley 2⅝ oz. and 60 grs., is therefore very considerable, and shows how much the zemindars would gain by growing wheat-barley and abandoning the cultivation of the degenerated husk-barley. The introduction of this cereal alone, on a suitable scale, would justify the formation of the much-needed seed farms.

WHEAT BARLEY.

DOORA MAIZE, AND SUGAR SORGHUM.

THE Kowur, Jainarain Sing, of Didwary, in the zillah or district of Meerut, reports that the *Doorra*, or many-branching variety of *Jowar*, has been successfully grown and developed heads of grain. The fourteen kinds of seed maize (mentioned in the chapter) sent him germinated, grew apace, and all bore cobs which came to maturity at the same time as the ordinary maize of the country and locality. The seed has been preserved for sowing in 1883.

This experiment is of very considerable value. The Kowur is an educated Talookdar, and a noble, and fully understands the value of improved agricultural products. The success attained shows that Moozuffurnuggur, fifty miles north of Meerut, is the proper place for a seed farm in the North-Western Provinces, and for this reason: if fourteen varieties of American maize have answered from thirty to forty miles east of Meerut, they and other kinds would grow very much better under the latitude of Moozuffurnuggur, which is also noted for its superior wheat crops. Canal irrigation is available, the soil is rich, and the line of railway runs through the station. The maize *Doorra*, and European wheats and barleys, as also wheat-barley acclimatized at this station, would suit all Rohilcund and the northern districts of Oudh. The imported wheats and barleys should in the first instance be grown in the Himalayas (Mussoorie being barred), and this seed be sent to Moozuffurnuggur to be sown in due course, and the acclimatized seed so obtained would then suit the localities named.

The *Doorra* was sown at Rajpore, and attained the height of fifteen feet; when the writer passed through a month ago,[1] the grain cobs had not been developed. Mr. P. Sri Lall, of Bijnor, was also supplied with some of the seed, but his report is not as yet to hand. The fact of this wonderfully productive cereal having matured seed east of Meerut shows that its extension is only a question of time; but, unlike American maize, the *Doorra* does not require to be acclimatized in the Himalayas.

[1] October 26, 1882.

SORGHUM SACCHARATUM (SUGAR SORGHUM).

IN Madras, during the season of 1877–78, Mr. Robertson, in charge of the Sydapet Government Model Farm, tried various experiments on guano, carbonate of lime, saltpetre, bone-dust, and sulphate of lime, to ascertain their value as manures.

The crop experimented on was the *Sorghum saccharatum* (Imphee). The guano gave 540 lbs. of grain and 2,808 lbs. of straw; value of increase per acre, Rs. 24. 4. 3. The saltpetre (nitrate of potash) gave an increase of 900 lbs. of grain and 7,380 lbs. of straw; increase per acre, Rs. 46. 7. 6. The profit, in the first instance is represented by Rs. 6. 4. 3, and in the second by Rs. 28. 76.. Rs. 18 per acre was expended on each description of manure, and Mr. Robertson states: "It is gratifying to find that saltpetre, an indigenous product, has given the best results; an expenditure of Rs. 18 per acre on this manure having not only nearly doubled the crop, but, after repaying its own cost, left an additional farmer's profit of nearly Rs. 30 per acre."

The above is taken from an official document, and fully supports the writer's views and experience on the value of the nitrate of potash as a mineral manure. In India, wherever nitrous soils exist, the nitre can be prepared therefrom at a prime cost of two rupees[1] (3s. 4d.) per maund of 82 lbs. At this rate the prime cost of 112 lbs., or 1 cwt., of saltpetre is 4s. 7d. But as an Allwise Being has associated saltpetre with common salt, the Excise Department claims and levies the full amount of salt duty on the salt, which the saltpetre maker does not want but must separate from the saltpetre; and, in addition, he has to pay the Excise duty levied on the nitre. But this is not all, for the man has to take out a license to manufacture saltpetre. The Excise Department does not know that the most valuable of mineral manures for wheat crops is saltpetre associated with common salt, exactly as it is found *in situ*. Zeal

[1] The information is derived from an official letter in the writer's possession.

without knowledge does not pay, and the above is a very good example thereof. The proper plan of procedure would be to free the manufacture from all restrictions, and to encourage the preparation of crude saltpetre and salt in combination, to be used for manurial purposes; the article as prepared being stored in bond, credit being given to the proprietor, who may sell at discretion to zemindars.

If at any time a demand for pure saltpetre arises, the owner may take out *quantum suff.* of the crude salts from bond, and manufacture the saltpetre, but without refining the common salt, which by a very simple catalytic process (at present unknown to the Excise and Salt Departments) may be rendered quite unfit for human food, though just the thing needed for cattle, by being converted into the chloride of potash, which is one of the constituents of cows milk, and is generally present in all plants growing in salt marshes.

If by the use of the manure named the yield of grain is doubled, as I venture to predict it will be (fossil phosphate of lime being also present), the Government might with justice claim a share of the extra produce, taking it in kind, but not its value in coin. The Government having by its act enabled the zemindar to obtain the manure at a cost fully within his reach and means, is fairly entitled to a share in the extra yield.

JAPAN PEA.

A SMALL supply of this pea-bean was received by the writer in the spring of 1882. It was freely distributed in Dehra, but failed entirely as a hot-weather crop. A few seeds sown at Mussoorie grew to the height of two feet, and bore pods very much resembling those of the *Urhur Dall* (*Cajanus sativa*). At the writer's request a supply of the seed was sent to Simla, and the result is given beneath.

"The Government of India has decided on the cultivation of the Japan pea being extended in this country, and with this view it has suggested that further experiments should be made in suitable places. This bean or pea has its natural habitat in China and Japan, it also grows in Mongolia, and in India in the Himalayas, and within the last few years it has been cultivated experimentally in several European countries, under the name of the Soy bean. It is a vegetable, which approaches most nearly in its proximate chemical composition to animal food. The Soy bean first attracted attention in Europe in the year 1873, when specimens from Japan, from China, and from India were shown at the Vienna International Exhibition. Dr. Forbes Watson, reporter on the products of India, called attention to it in the catalogue of the exhibits of the India Museum. Since then numerous experiments have been made on the European continent on its growth, and also feeding experiments with the bean and its straw on different kinds of animals have been prosecuted. Such experiments have been carried on by Woolling and Wein at Munich; by Haberlandt, Lehman, Harz, Stahel, Zimmerman, Siewert, Wieske, and others at various stations in Germany, Austria, and Hungary; and experiments have also been made in France and in Italy.

"In Japan it is known by names signifying THE bean, and from it are made not only soy but a paste, known as *miso*, which is in constant request at nearly every meal, *tofu* or bean cheese, and other foods used to a less extent. This bean cheese is also well known in China, and is obtained by extracting the legumin from the beans with water and precipitating it with brine. These foods are most valuable additions to the dietary of the Oriental nations, and especially of the Japanese, who use so little animal food. They tend to supply the deficiencies of the staple food, rice, in nitrogenous matter, fat, and also in mineral constituents. The Buddist priests, who are strictly forbidden the use of animal food, consume considerable quantities of these beans, principally in the form of *miso*."—*Vide* "*Simla Argus*," 18*th November*, 1882.

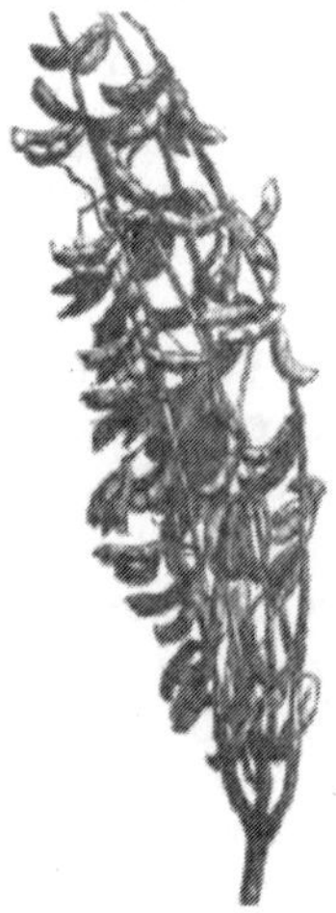

SOYA BEAN (SOJA HISPIDA), OR OLEAGINOUS JAPANESE PEA.

JAPAN RICE.

THE introduction into India of the upland, or dry rice of Japan, is a desideratum. When acclimatized, say in the Province of Chota Nagpore, at Poorulleea, and Hazareebaug, the cultivation could and would extend to the Monghyr, and the Rajmahal hills, and in like manner if acclimatized below Kotgurh, it would speedily spread through the various hill states lying on each side of the valley of the Sutledge, and confer a boon and blessing on all who cannot now grow rice, and who are looked down upon by all hill zemindars who own *Kyars*, or lands whereon wet rice is grown. In the Himalayas to own a rice field is a mark of respectability and wealth, and as this dry rice could be cultivated far and wide in the mountain valleys of the Punjab rivers, its introduction would enrich a very industrious but impoverished agricultural population.

In all the hilly districts of the Punjab, where from want of water-springs wet rice can never be grown, the upland dry rice of Japan could be most successfully cultivated. But, as before, nothing can be done without the seed farm. The Cawnpore Experimental Farm can no more grow dry rice than it can Ribston pippins, and it is to be hoped will not try the cultivation, which from want of information might be attempted.

RICE CULTIVATION IN JAPAN.

OF rice cultivation in Japan, Consul-General Van Buren tells us that—

"There are two general divisions of the rice plant (*Oriza sativa*) —'upland' and 'lowland.' The great bulk of the rice product is lowland rice, but the upland variety is not grown in all Asiatic countries. In some of the richest provinces of China, the tax or tribute collected in kind and sent to Peking for the Imperial use, consists, in great part, of upland rice. The lowland variety, in all cases, requires a low level soil, susceptible of being flooded several times during the season. The labour required for its production is immense. The plot of ground must be embanked, so as to hold the water, and the soil, after being flooded and exposed to the sun, bakes, and is worked with great difficulty.

"On the other hand, the upland rice is grown on high, dry ground, and in ordinary climates requiring no irrigation. The dry soil is easily and cheaply prepared for the seed, and needs no expensive system of irrigating ditches and embankments. The plot of ground can be so large as to admit of the employment of the plough, and the loose dry soil is fitted for its use. It is safe to say that the labour in Japan of producing an acre of upland rice is less than one-half that required for the lowland, and it may not be more than one-third. It is grown in all the *ken*, or districts of Japan, and flourishes in any soil adapted to wheat or barley. The soil is ploughed with the small Chinese plough drawn by one animal, usually a cow or a bull, or it is dug up with a mattock. The seed is sown in April or May, in drills about eighteen or twenty inches apart. In the drills, before the seed is sown, a compost of decomposed straw, closet manure, and ashes is strewn, upon which the seed is dropped, one to one and a quarter bushels per acre. The soil is dug up between the drills three or four times, to keep it loose and so destroy weeds. Two or three times during the growth of the plant small quantities of liquid fertilizers are poured on to the ground at its root. The ordinary height of the stalk when matured is about that of wheat or barley, but when the soil is very fertile, or an extra quantity of fertilizers has been used, it will sometimes reach a height of four and even five feet. It is ready for harvest in September or October, and is cut here with a sickle or knife, and threshed the same as wheat or barley.

"The process of hulling is the same as that used for lowland rice.

"An acre of land, which will produce a good crop of wheat or barley, will produce thirty bushels of upland rice.

"I have seen many acres yielding each 40 bushels or more. The weight of a bushel of this hulled rice is from 60 to 61½ lbs.

"The analysis of this rice, as given by Pavy, Edward Smith, and Parkes, is :—

"Nitrogenous matter	7·55
Starch	88·65
Dextrine	1·00
Fatty matter	0·80
Cellulose	1·10
Mineral matter	0·90
	100·00"

The above is taken from the *Indian Agriculturist* for September, 1882.

FODDER PLANTS.

MR. DUTHIE, F.L.S., Superintendent of the Government Botanical Gardens at Saharunpore, North-West Province, has during the season of 1881–82 made experiments on the culture of fodder plants, and in his report it is stated: "The experiments made with various fodder plants were valuable, as they attempted to deal with a weak spot in our agricultural economy.

"The greatest success was attained with guinea grass (*Panicum jumentorum*), which possesses three qualities making it valuable to the cultivators: it is a good feeding crop, it produces heavily, and the cultivation costs little. The outturn was at the rate of 600 maunds (49,000 lbs., or 21 tons 19¼ cwt.) per acre, and the cost of raising this may be considered almost *nil.* It must be freely watered to produce at this rate, and this seems to be almost the only expense connected with its cultivation."

In Madras the *Reana luxurians* was experimentally cultivated in the Chingleput district, and gave a return of 20 tons per acre, with an expenditure in seed of only 2 lbs. (*Vide* Mr. Robertson's Report for 1881.)

AGRICULTURAL WEIGHTS AND MEASURES.

IN India everything is bought and sold by the seer, which represents the weight of 80 rupees; the bushel is quite unknown, and Europeans accustomed to compute the yield of agricultural products by the bushel are often inconvenienced from want of information on this subject. The table given beneath supplies the want.

TABLE.

One Bushel of		lbs.	One Bushel of		lbs.
Wheat	weighs	60	Hemp seed	weighs	44
Shelled Indian corn ...	,,	56	Clover seed	,,	60
Rye	,,	56	Timothy grass seed ...	,,	45
Oats (common)	,,	32	Blue grass	,,	14
Barley ,, ...	,,	46	Dried peaches	,,	33
Buckwheat	,,	56	Rape seed	,,	48
Irish potatoes	,,	60	Peas	,,	64
Sweet potatoes	,,	60	A quarter of corn is 8 bushels, or		480
Onions	,,	57	A gallon of flour is		7
Beans	,,	60			
Bran	,,	27			

OIL MEASURES.

		lbs.			lbs.
1 Gallon	Olive oil is ..	7	1 Gallon	Sperm oil is ...	6¾
1 ,,	Castor oil is ...	7½	1 ,,	Paraffine oil is ...	6⅞
1 ,,	Whale oil is ...	7¼			

FIBRE-PRODUCING PLANTS.

THE zemindar does not know that a very considerable demand exists for fibres, which may at discretion be made into twine, and then woven into canvas and gunnybag material, the ends or cuttings being of value to the paper manufacturer.

The American aloe, now so commonly met with in most civil and military stations in the Bengal Presidency, possesses fine long leaves, which under suitable manipulation yield a strong and valuable fibre, and as the plant requires no care, its introduction on a large scale is desirable. If plants were forthcoming, every village could find room for their growth, either as a hedge plant, or for putting down on land unfit for the culture of grain crops owing to the preponderance of clay.

It cannot be too widely made known that the envelopes which cover the grain spike of all varieties of maize, and which are now thrown away or burned, afford a very valuable material for conversion into paper of the best description, such as is required for bank notes.

INDIAN OIL-PRODUCING TREES AND SHRUBS.

IN June, 1876, the writer brought the oil-producing trees and shrubs of British Burmah to public notice, through the columns of the *Delhi Gazette* newspaper, the object in view being their introduction into Bengal Proper. But up to date nothing has been done, and this great source of wealth remains neglected.

"The sweet or fatty oil-producing trees, sub-trees, and shrubs of British Burmah are as valuable as the *olive*, those producing naturally sweet-scented oils being more so. But there being no official system of organization for turning these great and permanent sources of wealth to account, they have remained *in statu quo*. As this paper is intended to draw the attention of the British capitalist to the natural resources of India, it may not be out of place to supply detailed information on the subject of these *oil-producers*.

"Of these the most valuable are—1. The *Connarus speciosa.* A large tree, very plentiful and remarkable for the quantity of its seeds, which are of large size, abounding in sweet oil.

"2. The *Connarus nitida*, a sub-tree about ten feet in height, very plentiful, and affords an oil seed of smaller size, but equally rich in a similar sweet oil.

"3. The *Galedupha arborea* and *Galedupha tetrapetala.* Both of these yield an oil expressed from the seed. Both trees are very common. The seed is large, and, it is alleged, might be collected in any quantity. The oil is commonly used for burning, and medicinally as an embrocation.

"4 and 5. The *Buchanania latifolia* and *Buchanania angustifolia.* Both trees are freely met with, and both yield in abundance valuable oil seeds of the sweet or fatty class.

"6. The *Xanthoxylon badrunga.* 'This plant affords a plentiful supply of oil seeds, which,' states Dr. J. McClelland, 'has not as yet, I believe, been taken advantage of as it deserves.'

"7. The *Calophylum longifolia* is, from its size, reckoned amongst timber trees, and affords an oil seed. This tree, together with the *Calophylum inophyllum* and *Calophylum lanceolaria*, bears fragrant flowers. Their seeds are large, and contain a considerable portion of oily matter which possesses much of the aroma of the flower.

"8. The *Spondias mangifera* is a tree which grows to a considerable size, and yields abundant seed rich in oil.

"The reader will hardly believe that this list of oil seeds 'might be extended,' and that it has purposely been confined to those which could be supplied in bulk in the Rangoon, Pegu, and Tounghoo districts of British Burmah."

The above was taken from an official report, printed for submission to the late Marquis Dalhousie, the great Governor-General of India.

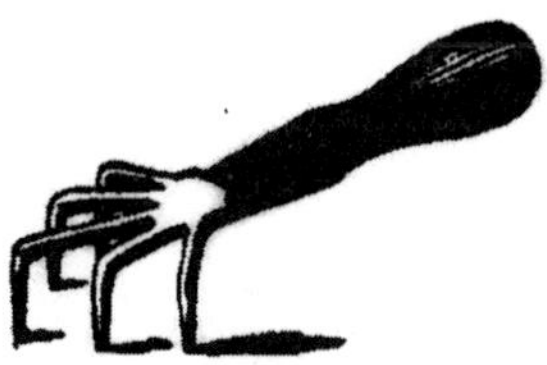

EXCELSIOR WEEDER.

INDEX.

THE END.

UNWIN BROTHERS, THE GRESHAM PRESS, CHILWORTH AND LONDON.

5 AND 6, GOVERNMENT PLACE, CALCUTTA.

A CATALOGUE

OF

THACKER, SPINK & CO.'S

PUBLICATIONS.

March, 1883.

5 AND 6, GOVERNMENT PLACE, CALCUTTA.

CONTENTS.

GENERAL INDEX.

A LIST OF

WORKS IN GENERAL LITERATURE

PUBLISHED OR SOLD BY

THACKER, SPINK AND CO.,

5 & 6, Government Place, Calcutta.

HISTORY, BIOGRAPHY, POLITICS, &c.

Bellew.—THE RACES OF AFGHANISTAN; being a Brief Account of the principal Nations inhabiting that Country. By Surgeon-Major H. W. BELLEW, C.S.I., late on special Political Duty at Cabul. 8vo., cloth. Rs. 3-8.

Aberigh-Mackay.—THE SOVEREIGN PRINCES AND CHIEFS OF CENTRAL INDIA. By G. R. ABERIGH-MACKAY, Principal, Residency (Rajcoomar) College, Indore, Central India. Illustrated with Portraits and Views. Volume I. Royal 8vo., cloth, extra gilt, and gilt top. Rs. 12.

NOTES ON WESTERN TURKISTAN. Some Notes on the Situation of Western Turkistan. By G. R. ABERIGH-MACKAY. 8vo. Rs. 2-8.

McCrindle.—ANCIENT INDIA AS DESCRIBED BY MEGASTHENES AND ARRIAN; being a Translation of the fragments of the "Indica of Megasthenes," collected by DR. SCHWANBECK; and a Translation of the first part of the "Indica of Arrian," with Introduction, Notes, and a Map of Ancient India. By J. W. MCCRINDLE, M.A., Principal of Patna College. 8vo. Rs. 2-8.

THE COMMERCE AND NAVIGATION OF THE ERYTHRÆAN SEA; being a Translation of Periplus Maris Erythræi by an Anonymous Writer, and of Arrian's Account of the Voyage of Nearkhos from the mouth of the Indus to the head of the Persian Gulf. With Introduction, Commentary, Notes, and Index. By J. W. MCCRINDLE, M.A., Principal of the Government College, Patna. 8vo. Rs. 3.

Wilkins.—A HANDBOOK OF HINDU MYTHOLOGY. Vedic and Puranic. By Rev. W. J. WILKINS, of the London Missionary Society, Calcutta. With 70 Illustrations of Hindu Deities from Native Drawings, Square, uniform with "Lays of Ind," and Hayes's "Riding on the Flat and Across Country." Rs. 7.

Sherring.—HINDU TRIBES AND CASTES. Vol. III. Together with three Dissertations: on the Natural History of Hindu Caste; the Unity of the Hindu Race; and the Prospects of Indian Caste. Including a General Index of the three Volumes. By the Rev. M. A. SHERRING, M.A., LL.B. Demy 4to., cloth. Rs. 16.

This Volume contains the Tribes and Castes of Rajputana, the Madras Presidency, Mysore, Travancore, &c.

Baness.—INDEX GEOGRAPHICUS INDICUS: being a Statistical, Political, and Descriptive Account of the several Provinces and Administrations of the Empire, and the Independent, Feudatory, and outlying Native States; and the Foreign and Colonial Possessions in the East. Together with a Glossary, Statistical Tables, the Religions and Peoples of India, and an Alphabetical List of the Principal Places in Her Majesty's Indian Empire. With 8 Maps, drawn on a scale of 64 miles=1 inch. By J. FREDERICK BANESS, F.R.G.S. Royal 8vo., half-bound. Rs. 12; post free 12-8.

Wheeler.—TALES FROM INDIAN HISTORY: being the Annals of India retold in Narrative. By J. TALBOYS WHEELER. Rs. 3-8.

Wheeler.—EARLY RECORDS OF BRITISH INDIA, a History of Early English Settlements as told in Government Records, old Travels, and contemporary documents down to the rise of British Power in India. By J. TALBOYS WHEELER. Royal 8vo., cloth. Rs. 5.

Sherring.—HINDU TRIBES AND CASTES; Vol. II. together with an Account of the Mahomedan Tribes of the North-West Frontier and of the Aboriginal Tribes of the Central Provinces. By the Rev. M. A. SHERRING, M.A., LL.B. Demy 4to., cloth. Rs. 16.

This Volume contains an account of the Tribes of the Punjab and its Frontiers, Central Provinces, and Berar; Bombay Presidency, and Frontiers of Scind

McMahon.—THE KARENS OF THE GOLDEN CHERSONESE. By Lt.-Col. A. R. MCMAHON, F.R.G.S. With Maps. Demy 8vo., cloth. Rs. 12-14.

Mead.—SEPOY REVOLT: its Causes and its Consequences. By H. MEAD. 8vo., cloth. Rs. 4.

Malcolm.—A MEMOIR OF CENTRAL INDIA, including Malwa and adjoining Provinces, with the History, and Copious Illustrations, of the past and present conditions of that Country. By Major-General Sir JOHN MALCOLM, G.C.B., K.L.S. Reprinted from the Third Edition. With Map. In 2 vols. crown 8vo. Rs. 7.

Wyllie.—ESSAYS ON THE EXTERNAL POLICY OF INDIA. By W. S. WYLLIE M.A. Edited, with a Memoir and Notes, by W. W. HUNTER, B.A., LL.D. With Portrait. 8vo. Rs. 10.

Rajendra Lala Mittra.—THE PARSIS OF BOMBAY: a Lecture delivered on February 26, 1880, at a Meeting of the Bethune Society. By RAJENDRA LALA MITTRA, LL.D., C.I.E. 8vo., sewed. Re. 1.

Hunter.—THE INDIAN MUSSULMANS. By Dr. W. W. HUNTER. Second Edition. 8vo. Rs. 7-8.

A LIFE OF THE EARL OF MAYO, fourth Viceroy of India. With a Narrative of his Indian Administration. By W. W. HUNTER. 2 vols. demy 8vo. Rs. 17-2.

A STATISTICAL ACCOUNT OF BENGAL. By W. W. HUNTER. 20 vols. 8vo., half bound. Price per volume, Rs. 4.

ANNALS OF RURAL BENGAL. By W. W. HUNTER. Fifth Edition. 8vo., cloth. Rs. 12 4.

ORISSA; or, the Vicissitudes of an Indian Province under Native and British Rule. With Engravings and Map. By W. W. HUNTER. 2 vols. demy 8vo., cloth. Rs. 22-8.

Briggs.—THE NIZAM: his History and Relations with the British Government. By HENRY GEORGE BRIGGS. With Portrait. 2 vols 8vo. Rs. 15.

Thorburn.—BANNU; or, Our Afghan Frontier. By S. S. THORBURN, C.S., Settlement Officer. With Map. 8vo., cloth. Rs. 12-14.

Loke Nath Ghose—THE INDIAN PEERAGE: Modern History of the Indian Chiefs, Rajas, Zemindars, &c. By BABU LOKE NATH GHOSE. 2 vols. 8vo., cloth. Rs. 10; post-free 10-8.

Laurie.—SKETCHES OF SOME DISTINGUISHED ANGLO-INDIANS, with an account of Anglo-Indian Periodical Literature By Col. W. F. B. LAURIE. With Portrait of Sir John Kaye. 12mo., cloth. Rs. 3-6.

The Rajatarangini.—KINGS OF KASHMIRA; being a Translation of the Sanskrit Work, Rajatarangini of Kahlana Pandita. By JOGESH-CHUNDER DUTT. Small 8vo. Rs. 2.

Smith.—MOHAMMED AND MOHAMMEDANISM: Lectures delivered at the Royal Institution of Great Britain in February and March, 1874. By R. BOSWORTH SMITH. Second Edition, revised, with considerable additions. 8vo., cloth. Rs. 6-2.

Kally Prosono Dey—SIR LOUIS CAVAGNARI, Life and Career of. Together with a brief outline of the 2nd Afghan War. By BABU KALLY PROSONO DEY. 8vo., cloth. Rs. 5: post-free 5-4.

Muir.—THE LIFE OF MAHOMET, from Original Sources. By Sir WILLIAM MUIR. New and cheaper Edition, in one vol. With Maps. Demy 8vo., cloth. Rs. 10.

The Original Edition, in four volumes 8vo. Rs. 30.

Haggard.—ENGLAND AND ISLAM. By A. H. HAGGARD, B.C.S. 8vo. Re. 1.

Syed Ameer Ali.—A CRITICAL EXAMINATION OF THE LIFE AND TEACHINGS OF MOHAMMED. By SYED AMEER ALI. Post 8vo., cloth. Rs. 6-8.

CENTRAL ASIAN QUESTIONS FROM AN EASTERN STAND-POINT. 8vo., stitched. Rs. 1-8.

Haughton.—CHAREEKAR and Service there with the 4th Goorkha Regiment (Shah Shooja's Force), in 1841. An Episode of the First Afghan War. By Col. HAUGHTON, C.S.I. 8vo., sewed. As. 8.

Outram.—LIFE OF LIEUTENANT-GENERAL SIR JAMES OUTRAM. By Major-General Sir FREDERIC J. GOLDSMID, C.B., K.C.S.I. With Illustrations and Maps. 2 vols. 8vo. Rs. 23.

Munro (Major-Gen. Sir Thomas, Bart., K.C.B., Governor of Madras). SELECTIONS FROM HIS MINUTES AND OTHER OFFICIAL WRITINGS. Edited, with an Introductory Memoir, by Sir ALEXANDER ARBUTHNOT, K.C.S.I., C.I.E. 2 vols. Demy 8vo., cloth, price Rs. 21.

TOPOGRAPHY, &c.

Keene.—A HANDBOOK FOR VISITORS TO AGRA AND ITS NEIGHBOURHOOD. By H. G. KEENE, C.S. Fourth Edition, revised. Maps, Plans, &c. Fcap. 8vo., cloth. Rs. 2-8.

A HANDBOOK FOR VISITORS TO DELHI AND ITS NEIGHBOURHOOD. By H. G. KEENE, C.S. Third Edition. Maps Fcap. 8vo., cloth. Rs. 2.

A HANDBOOK FOR VISITORS TO ALLAHABAD, CAWNPORE, AND LUCKNOW. By H. G. KEENE, C.S. Second Editon, re-written and enlarged. Fcap. 8vo., Rs. 2-8,

Ince.—THE KASHMIR HANDBOOK; a Guide for Visitors. With Map and Routes. By JOHN INCE, M.D. Third Edition. Rs. 5.

CALCUTTA TO LIVERPOOL, BY CHINA, JAPAN, AND AMERICA, IN 1877. By Lieut. General Sir HENRY NORMAN. Second Edition. Fcap. 8vo., cloth. Rs. 2-8.

HOW WE DID "THE LIONS" OF THE NORTH-WEST: a Trip in the Durga-Pujahs to Lucknow, Delhi, Agra. By F. O. B. Re. 1.

Bholanath Chunder.—TRAVELS OF A HINDU TO VARIOUS PARTS OF BENGAL AND UPPER INDIA. By BHOLANATH CHUNDER. With an Introduction by J. TALBOYS WHEELER. 2 vols. small 8vo., cloth. Rs. 12.

Busteed.—ECHOES FROM OLD CALCUTTA; being chiefly Reminiscences of the days of Warren Hastings, Francis, and Impey. By H. E. BUSTEED. Rs. 4.

Stephen.—THE DELHI HANDBOOK. By CARR-STEPHEN. Stitched. Rs. 8.

THE ARCHÆOLOGY AND MONUMENTAL REMAINS OF DELHI. By CARR-STEPHEN. Royal 8vo., cloth. Rs. 10.

Villeroi.—A TRIP THROUGH CENTRAL ASIA. By BERNARD DE VILLEROI. 12mo. Re. 1.

KASHGARIA (Eastern or Chinese Turkestan). Historical and Geographical Sketch. Military Strength, Industries and Trade. By Col. A. N. KUROPATKIN, Russian Army. Translated by W. E. GOWAN, H.M.'s. Indian Army. Rs. 6.

Cornish. — UNDER THE SOUTHERN CROSS. By HENRY CORNISH. Second Edition. Enlarged and illustrated. 8vo. Rs. 3.

Vaughan.—THE MANNERS AND CUSTOMS OF THE CHINESE OF THE STRAITS SETTLEMENTS By J. D. VAUGHAN, Barrister-at-Law. 8vo., stiff cover. Rs. 3.

AN ACCOUNT OF THREE MONTHS' TOUR FROM SIMLA, through Bussahir, Kunowar, and Spiti, to Lahoul (In the Footsteps of the Few). By Mrs. J. C. MURRAY-AYNSLEY.

A GUIDE TO MUSSOOREE, LANDOUR, DEHRA DUN, AND THE HILLS NORTH OF DEHRA, including Routes to the Snows and Places of Note. With Chapters on Hurdwar, Roorkee, &c. By J. NORTHAM.

NATURAL AND PHYSICAL SCIENCES.

Beddome.—HANDBOOK TO THE FERNS OF BRITISH INDIA, CEYLON, AND THE MALAY PENINSULA. With 300 Illustrations. By Col. R. T. BEDDOME. Imperial 16mo.

Sterndale.—NATURAL HISTORY OF THE MAMMALIA OF INDIA. With numerous Illustrations. By R. A. STERNDALE. Imperial 16mo. Rs. 7. (*In the press.*)

Clarke.—COMPOSITÆ INDICÆ DESCRIPTÆ ET SECUS GENERA BENTHAMII ORDINATÆ. By C. B. CLARKE, M.A. Demy 8vo. Rs. 1-8.

ROXBURGH'S FLORA INDICA; or, Descriptions of Indian Plants. Reprinted literatim from Carey's Edition. 8vo., cloth. Rs. 5.

Gregg.—A TEXT-BOOK OF INDIAN BOTANY; Morphological, Physiological, and Systematic. Profusely illustrated. By W. H. GREGG, B.M.S., Lecturer on Botany at the Hugli Government College. Chapters I. and II. in part. Crown 8vo. Re. 1. (The completion of the work is in active preparation.)

Day.—REPORT ON THE FRESH WATER FISH AND FISHERIES OF INDIA AND BURMAH. By J. DAY. 8vo., boards. Rs. 2.

Kurz.—FOREST FLORA OF BRITISH BURMAH. By S. KURZ, Curator of the Herbarium, Royal Botanical Gardens, Calcutta. 2 vols. demy 8vo., cloth. Rs. 15.

Beavan.—HANDBOOK OF THE FRESH WATER FISHES OF INDIA. Giving the characteristic peculiarities of all the Species at present known, and intended as a guide to Students and District Officers. By Capt. R. BEAVAN, F.R.G.S. Twelve full Plates. Demy 8vo. Rs. 7-8.

Theobald.—DESCRIPTIVE CATALOGUE OF THE REPTILES OF BRITISH INDIA. By W. THEOBALD, Geographical Survey. 8vo., cloth. Rs. 10.

Nevill.—A CATALOGUE OF MOLLUSCA IN THE INDIAN MUSEUM. By G. NEVILL. Demy 8vo., stitched. Rs. 2.

Dobson.—A MONOGRAPH OF THE ASIATIC CHIROPTERA AND CATALOGUE OF BATS IN THE COLLECTION OF THE INDIAN MUSEUM, CALCUTTA. By G. S. DOBSON. Demy 8vo., cloth. Rs. 6.

Gore.—A GLOSSARY OF FOSSIL MAMMALIA. For the use of Students of Palæontology. By J. E. GORE. Fcap. 8vo., cloth. Rs. 1-8

SOUTHERN STELLAR OBJECTS between the Equator and 55° South Declination. By J. E. GORE. 8vo., cloth. Rs. 2-8.

Jerdon.—THE BIRDS OF INDIA: being a Natural History of all the Birds known to inhabit continental India, a Manual of Ornithology specially adapted for India. By the late T. C. JERDON, to which is added his Supplementary Notes published in the "Ibis." 3 vols. 8vo., cloth. Rs. 30.

Blanford.—THE INDIAN METEOROLOGIST'S VADE MECUM. Part I.—Instructions to Meteorological Observers in India. Part II.—Meteorology of India. With Tables for the reduction of Meteorological Observations in India. By H. F. BLANFORD. 2 parts, large 8vo., cloth. Rs. 10.

RAINFALL TABLES. Table of Average Monthly and Annual Rainfall in Northern India, &c. One Sheet. As. 4.

McGregor.—PROTECTION OF LIFE AND PROPERTY FROM LIGHTNING DURING THUNDER-STORMS. By W. McGREGOR. Second Edition, enlarged. Stitched. Re. 1.

Archibald.—THE RAINFALL OF THE WORLD in connection with the eleven-years period of Sunspots. With an Introduction and Appendix. By E. D. ARCHIBALD, B.A., F.M.S. Demy 8vo., stitched. Rs. 2.

Elson.—THE SANDHEADS SAILING DIRECTORY; or, The Sub-Marine Hooghly Delta; its Marks and Deeps, its Winds, Weather, Tides, and Currents, being an Epitome of useful information for Ship-masters, also a Handy Book of Reference for Pilots. By SAMUEL R. ELSON, Senior Master Pilot, Author of the "Morse Universal Marine Flashing Signal," "The Sailors' East Indian Sky Interpreter," "Circular Diagram for Compass Deviation" and the "Bengal Pilot's Code of Signals." Rs. 8.

THE SAILOR'S EAST INDIAN SKY INTERPRETER AND WEATHER BOOK; being a Description of the Phenomena and Prognostics of the Bay of Bengal October-Cyclones, as experienced at the Pilot Station off the mouth of the Hooghly. By S. R. ELSON. 8vo. Re. 1.

THE MORSE UNIVERSAL MARINE FLASHING SIGNALS, for use, with the Commercial Code Signal Book in foggy weather, Night Signalling, &c., with a selection of twenty-one handy Night Towing Signals. By S. R. ELSON, Licensed Hooghly Pilot. 8vo., sewed. As. 12.

Yonge.—LATITUDE AND LONGITUDE WITHOUT INSTRUMENTS; being a Chart for finding the Hour of Sunrise and Sunset for every day in the year, at every place in the Northern or Southern Hemisphere. By Capt. W. L. YONGE, R.A. On paper, Re. 1; on India-rubber, Rs. 2-8.

FINANCE, PRODUCE, &c., OF INDIA.

Seyd.—THE FALL IN THE PRICE OF SILVER; its Causes, its Consequences, and their possible Avoidance, with special reference to India. By ERNEST SEYD, F.S.S. 8vo. Rs. 1-12.

Strachey—THE FINANCES AND PUBLIC WORKS OF BRITISH INDIA, from 1869 to 1881. By Sir JOHN STRACHEY, G.C.S.I., and Lieut.-Gen. RICHARD STRACHEY, R.E., F.R.S. Demy 8vo. Rs. 12-8.

TEA GARDENS, COFFEE PLANTATIONS, INDIGO, JUTE, LAC, AND CINCHONA CONCERNS IN INDIA, with their Capital, Directors, Proprietors, Agents, Managers, &c., and their Factory Marks. Rs. 2-8.

Hector.—CURRENCY CONSIDERED WITH SPECIAL REFERENCE TO THE FALL IN THE VALUE OF SILVER AND ITS CONSEQUENCES TO INDIA. By J. HECTOR. 8vo. Pamphlet. Rs. 1-6.

Smeaton.—THE CURRENCY OF INDIA: an Examination of the reasons for change of the present Standard of Value and of the mode of effecting it. By D. MACKENZIE SMEATON, Bengal Civil Service. Demy 8vo., stitched. Re. 1.

Oakeley.—REMARKS ON MR. WILSON'S MINUTE ON THE INTRODUCTION OF A GOLD CURRENCY IN INDIA. By H. E. OAKELEY, M.A. 8vo. Pamphlet. Re. 1.

Lees.—THE DRAIN OF SILVER TO THE EAST, AND THE CURRENCY OF INDIA. By Colonel W. NASSAU LEES, LL.D. Demy 8vo., cloth. Rs. 5.

MEMORANDUM WRITTEN AFTER A TOUR THROUGH THE TEA DISTRICTS IN EASTERN BENGAL IN 1864–65. By Colonel W. NASSAU LEES, LL.D. Rs. 5.

THE LAND AND LABOUR OF INDIA, a Review. By Colonel W. NASSAU LEES, LL.D. 8vo., cloth. Rs. 4.

Pogson.—MANUAL OF AGRICULTURE FOR INDIA. In the form of Question and Answer. By Lieut. FREDK. POGSON. Crown 8vo.

Baildon.—TEA IN ASSAM. A Pamphlet on the Origin, Culture, and Manufacture of Tea in Assam, with an Appendix—Rural Life amongst the Assamese. By S. BAILDON. Demy 8vo. Rs. 2.

Hem Chunder Kerr.—REPORT ON THE CULTIVATION OF AND TRADE IN, JUTE IN BENGAL, and on Indian Fibres available for the Manufacture of Paper. By HEM CHUNDER KERR. Fcap. folio. Paper boards. With Map. Rs. 8.

Hughes.—NOTES ON THE COST OF CARRIAGE BY RAIL AND WATER IN THE UNITED STATES OF AMERICA. With Maps and Statistical Plans. By A. J. HUGHES, C.E. Demy 8vo., boards. Rs. 3-8.

Stoker.—MANAGEMENT OF THE TEA PLANT. By T. STOKER. 8vo., stitched. Rs. 2.

Watson.—REPORTS ON COTTON GINS and on the Cleaning and Quality of Indian Cotton. By D. FORBES WATSON. Part II.—Detailed Description with Records of the Cotton Gin. Trials at Manchester, Dharwar, and Broach (published by order of the Secretary of State for India in Council). Imperial 8vo. Rs. 8.

THE INDIAN RYOT, LAND TAX, PERMANENT SETTLEMENT, AND THE FAMINE. Compiled by ABHAY CHARN DAS, a Ryot. 8vo., cloth. Rs. 4.

"An endeavour to point out the effects of the fiscal measures upon the poor tenantry from a ryot's point of view."

Digby.—FAMINE CAMPAIGN IN SOUTHERN INDIA (Madras and Bombay Presidencies and Province of Mysore), 1876–78. By WILLIAM DIGBY, Hony. Secy., Indian Famine Relief Fund. Two vols. Demy 8vo., cloth. Rs. 16.

Money.—CULTIVATION AND MANUFACTURE OF TEA. By Lieut.-Col. EDWARD MONEY. Prize Essay. Third Edition, corrected and much enlarged. Demy 8vo., cloth. Rs. 5-6.

Hull.—COFFEE PLANTING IN SOUTHERN INDIA AND CEYLON; being a Second Edition, revised and enlarged, of "Coffee," its Physiology, History, and Cultivation. By E. C. P. HULL. 12mo., cloth. Rs. 6.

Schrottky.—MAN, PLANT, AND SOIL, AND THEIR CORRELATIONS. By EUGENE C. SCHROTTKY. 8vo., paper. Re. 1.

MILITARY WORKS.

Barrow.—THE SEPOY OFFICER'S MANUAL; a Book of Reference for Officers of the Bengal Native Infantry. By Lieut. E. G. BARROW, 7th N. I. Pocket. Rs. 2-8.

BENGAL ARMY REGULATIONS. Revised Army Regulations corrected to 1880. Regulations and Orders for the Army of the Bengal Presidency. 8vo., cloth. Rs. 4. Interleaved copies, Rs. 5.

BENGAL ARMY LIST. The Official Quarterly Army List of H.M.'s Forces in Bengal, exhibiting the Rank, Dates of Commission, and present Location of every Officer in the Bengal Army. Published January, April, July, and October in each year. Each Rs. 1.

Bell.—INSTUCTIONS IN CAVALRY DRILL. Part III.—Skirmishes, Scouts, Advance, Rear and Flank Guards, Outposts and Patrols. By C. W. BOWDLER BELL. Fcap. 8vo., stiff cover. Rs. 1-12.

Sterndale.—MUNICIPAL WORK IN INDIA; or, Hints on Sanitation, General Conservancy, and Improvement in Municipalities. Towns and Villages. By REGINALD CRAUFUIRD STERNDALE. Crown 8vo., cloth. Rs. 5. *(Shortly.*

PAY CODE. Revised Code of existing Orders regarding Pay, Pension, and Promotion of the Military Services in India. Corrected to 1st May, 1876. Royal 8vo., cloth.

Vol. I.—British Troops, including Officers and Men, Indian Army. Rs. 4.

Vol. II.—Native Troops, including Followers of British and Native Troops. Rs. 2.

Furse.—STUDIES ON MILITARY TRANSPORT. By G. A. FURSE, Major, 42nd Royal Highland Regiment. 8vo., stitched. Rs. 3.

Schmid.—THE SOLDIER IN THE FIELD. A Guide to the Theoretical Instruction of Non-Commissioned Officers and Soldiers in Picquet and Outpost Duties. By Major A. SCHMID. Sewed. Re. 1.

Stubbs.—HISTORY OF THE ORGANIZATION, EQUIPMENT, AND WAR SERVICES OF THE REGIMENT OF BENGAL ARTILLERY. Compiled from Published Works, Official Records, and various private sources, by Colonel FRANCIS W. STUBBS. With numerous Illustrations and Maps. Demy 8vo., cloth. 2 vols. Rs. 28.

Trench.—SHORT SERVICE AND DEFERRED PAY. By Captain F. CHENEVIX TRENCH. Second Edition, revised and enlarged. 8vo., stitched. Rs. 2-8.

Voyle.—A MILITARY DICTIONARY, comprising Terms, scientific and otherwise, conected with the Science of War. By Major-General G. E. VOYLE. Third Edition (London, 1876). 8vo. Rs. 7-8.

Broome.—THE RISE AND PROGRESS OF THE BENGAL ARMY. Maps and Plans. By General A. BROOME. Vol. I. (all published). 8vo., cloth. Rs. 5.

TARGET PRACTICE. The Soldiers' and Volunteers' Handbook. How to Shoot straight. 12mo. Rs. 2.

I. C. Bose—A HANDBOOK OF RULES RELATING TO PAY, Service Examinations, and Travelling Allowances of Officers in Civil Employ in the Lower Provinces. Compiled by I. C. BOSE, M.A. Rs. 3; post-free 3-5.

VETERINARY WORKS, SPORTS, &c.

Humfrey—THE STEEPLECHASE HORSE: how to Select, Train, and Ride Him. With Notes on Accidents and Diseases, and their Treatment. By Capt. J. HUMFREY. Cloth, limp. Rs. 3-8.

Hayes.—VETERINARY NOTES FOR HORSE-OWNERS. With Hindustani Vocabulary; an every-day book. By Capt. M. HORACE HAYES. Illustrated. Crown 8vo. Second Edition, enlarged. Rs. 7.

A GUIDE TO TRAINING AND HORSE MANAGEMENT IN INDIA. With a Hindustanee Stable and Veterinary Vocabulary and Calcutta Turf Club Tables of Weight for Age and Class. By Capt. M. HORACE HAYES, Author of "Veterinary Notes for Horse-Owners." New Edition, re-arranged and much enlarged. Crown 8vo. Rs. 5.

RIDING ON THE FLAT AND ACROSS COUNTRY. A Guide to Practical Horsemanship. By Captain M. HORACE HAYES. Handsomely illustrated and bound. Rs. 7.

INDIAN RACING REMINISCENCES. With 21 Portraits and other Illustrations. By Capt. M. H. HAYES.

Oriental Sporting Magazine—FROM JUNE 1828 TO JUNE 1833. In two volumes 8vo. Rs. 7-8; cash 6-12.

Sanderson.—THIRTEEN YEARS AMONG THE WILD BEASTS OF INDIA; their Haunts and Habits, from the Personal Observation: with an account of the Modes of Capturing and Taming Wild Elephants. By G. P. SANDERSON, Officer in Charge of the Government Elephant Keddahs at Mysore. With 21 full-page Illustrations and three Maps. Second Edition. Fcap. 4to. Rs.

Dogs.—INDIAN NOTES ABOUT DOGS; their Diseases and Treatment. Compiled by Major C. ——, author of "Horse Notes." Fcap 8vo., cloth. Re. 1-8.

Baldwin.—THE LARGE AND SMALL GAME OF BENGAL AND THE NORTH-WESTERN PROVINCES OF INDIA. By Capt. J. H. BALDWIN. With numerous illustrations. 4to., cloth. Rs. 15.

Kinloch. — LARGE GAME SHOOTING IN THIBET AND THE NORTH-WEST. By Capt. ALEX. A. KINLOCH, C.M.Z.S., 60th Royal Rifles. Illustraated by Photo. tint. New Edition. 2 vols. in 1. 4to. Illustrated. (*In the press.*)

DOMESTIC ECONOMY AND MEDICINE.

Birch. — GOODEVE'S HINTS FOR THE MANAGEMENT AND MEDICAL TREATMENT OF CHILDREN IN INDIA. By EDWARD A. BIRCH, M.D., Surgeon-Major. Seventh Edition. Crown 8vo., cloth. Rs. 7.

"I have no hesitation in saying that the present one is for many reasons superior to its predecessors. It is written very carefully, and with much knowledge and experience on the author's part, whilst it possesses the great advantage of bringing up the subject to the present level of Medical Science."—DR. GOODEVE.

GOODEVE'S HINTS FOR THE GENERAL MANAGEMENT OF CHILDREN IN INDIA in the absence of Professional Advice. Entirely re-written by JOSEPH EWART, M.D. Limp cloth. Rs. 2.

Indian Cookery Book: A Practical Handbook to the Kitchen in India, adapted to the three Presidencies; containing original and approved Recipes in every Department of Indian Cookery; Recipes for Summer Beverages and Home-made Liqueurs; Medicinal and other Recipes; together with a variety of things worth knowing. By A THIRTY-FIVE YEARS' RESIDENT. Crown 8vo. Rs. 3.

Riddell.—INDIAN DOMESTIC ECONOMY AND RECEIPT BOOK. With Hundustani Names. By DR. R. RIDDELL. Eighth Edition. Royal 12mo. Rs. 6.

Duke.—QUERIES AT A MESS TABLE: What shall we eat? What shall we drink? By JOSHUA DUKE, Surgeon, 3rd Punjab Cavalry, Author of "Banting in India." Fcap. 8vo., cloth, gilt. Rs. 2-4.

HOW TO GET THIN; or, Banting in India. By J. DUKE. 12mo., stiff cover. Re. 1.

Hogg.—PRACTICAL REMARKS CHIEFLY CONCERNING THE HEALTH AND AILMENT OF EUROPEAN FAMILIES IN INDIA, with special reference to Maternal Management and Domestic Economy. By FRANCIS R. HOGG, M.D., Surgeon-Major. Demy 8vo., boards. Rs. 2.

Hunter.—MEDICAL HINTS FOR THE DISTRICTS, AND COMPANION TO THE DISTRICT MEDICINE CHEST, arranged for Government Officials, their Families, and Camp-followers, and for all who are cut off from oral Medical Advice. By GEORGE YEATES HUNTER, Surgeon, Bombay Army. Fcap. 8vo., cloth. Rs. 3-8.

HEALTH IN INDIA; Medical Hints as to who should go there; how to retain Health there and on returning Home. By GEORGE YEATES HUNTER. 16mo., cloth, gilt. Rs. 2.

GARDENING.

Firminger.—A MANUAL OF GARDENING FOR BENGAL AND UPPER INDIA. By T. A. C. FIRMINGER, Honorary Member, Agri.-Horti. Society. Third Edition. 8vo., cloth, gilt. Rs. 10.

THE AMATEUR GARDENER IN THE HILLS: being Hints taken from various works by Authorities on Garden Management, and adapted to the Hills; to which are added a few Hints on Fowls, Pigeons, and Rabbit keeping; and various Recipes connected with the above subjects, which are not commonly found in Receipt Books. By AN AMATEUR. Crown 8vo. Rs. 3.

Graham.—THE CULTURE OF EXOTIC ROSES IN INDIA MADE EASY: Practical Directions for their improved Training, Propagation, and General Treatment. New Edition, revised by JAMES GRAHAM. Fcap. 8vo., cloth. Rs. 2-8.

Lowis.—FAMILIAR INDIAN FLOWERS. Thirty coloured plates and descriptive letterpress. By LENA LOWIS. 4to., cloth. Rs. 22-8.

MEDICAL WORKS.

Russell.—MALARIA; ITS COURSE AND EFFECTS. Injuries of the Spleen: an Analysis of Thirty-nine Cases. By E. G. RUSSELL, M.B., B.Sc., London. Demy 8vo., cloth gilt. Rs. 8.

"Discusses most exhaustively the subject of Malaria and its effects, and the work should be read by all who are liable to be called upon to treat the conditions to which it gives rise. . . . We leave Dr. Russell's volume commending it to the notice of our readers as a book full of facts and of clinical research, and which must be considered as an authority on the subject."—*The Lancet.*

"Besides the results of his personal observations, he summarises the views of recognised authorities on his subject, and thus presents to his readers a convenient epitome of previous investigations."—*Medical Press and Circular.*

Hall.—CAUSES OF BLINDNESS IN INDIA, Treatment and Cure, with an Appendix on Hypermetropia, Diagnosis, &c. By Surgeon GEOFFREY C. HALL, Indian Medical Service. Fcap. 8vo., cloth, gilt. Rs. 2.

Roy.—THE CAUSES, SYMPTOMS, AND TREATMENT OF BURDWAN FEVER: or, The Epidemic Fever of Lower Bengal. By GOPAUL CHUNDER ROY, M.D., F.R.C.S., London (Exam.) 8vo., cloth. Rs. 3-6.

Dutt.—THE MATERIA MEDICA OF THE HINDUS. Compiled from Sanscrit Medical Works by UDOY CHAND DUTT; with a Glossary of Indian Plants, by G. KING, M.B., F.L.S. Demy 8vo., cloth. Rs. 6.

Hume—JAIL HYGIENE IN INDIA. By T. HUME M.B. Fcap. 8vo., sewed. Rs. 2-8; post-free 2-10.

Hutchinson.—A GLOSSARY OF MEDICAL AND MEDICO-LEGAL TERMS, including those most frequently met with in the Law Courts. Compiled by R. F. HUTCHINSON, M.D., Surgeon-Major, Bengal Army. Fcap. 8vo., cloth. Rs. 3.

Ranking.—HIDAYAT-UL-HUKUMA. A Hindustani Manual for the use of Indian Medical Officers, Native Doctors, and others. By GEORGE S. A. RANKING, B.A., M.B., Cantab., Surgeon, 6th Bengal Cavalry. 18mo., sewed. Rs. 1-4.

Barnard.—CHOLERA MALIGNA is a Specific Acute Inflammation of the Mucous Tissue of the Small Intestines. By G. BARNARD. 8vo. Pamphlet. Re. 1.

Billing.—THE NATURE AND TREATMENT OF ASIATIC CHOLERA. By ARCHIBALD BILLING. 12mo. Re. 1.

THE INDIAN ANNALS OF MEDICAL SCIENCE. Nos. XXVIII. to XXXVII., January, 1877. Rs. 4 each.

Munro.—LEPROSY. By W. MUNRO, M.D., C.M. 8vo., boards. Rs. 3-4.

Chevers.—A PRACTICAL TREATISE ON THE MANAGEMENT OF THE DISEASE OF THE HEART AND OF AORTIC ANEURISM, with especial reference to the Treatment of those Diseases in India. By Dr. NORMAN CHEVERS. Demy 8vo., cloth. Rs. 2-8.

Bose.—PRINCIPLES OF RATIONAL THERAPEUTICS, commenced as an Inquiry into the relative value of Quinine and Arsenic in Ague. By BHOLANATH BOSE, M.D. Crown 8vo. Rs. 3-4.

A NEW SYSTEM OF MEDICINE, entitled Recognizant Medicine or, The State of the Sick. By BHOLANATH BOSE. 8vo. Rs. 7-8.

O'Shaughnessy.—THE BENGAL DISPENSATORY. Compiled chiefly from the works of Roxburgh, Wallich, Royle, Ainslie, Wight, Arnoll, Pereria, Fee, Richard, and Lindlay, including the Results of numerous Special Experiments. By W. B. O'SHAUGHNESSY. Large 8vo., cloth. Rs. 8.

Martin.—THE NATURE OF THE TISSUE-CHANGE KNOWN AS FATTY DEGENERATION OF THE HEART. By J. MARTIN, Medical Department. 8vo., cloth. Re. 1.

Playfair.—A TREATISE ON THE SCIENCE AND PRACTICE OF MIDWIFERY. By W. S. PLAYFAIR, M.D. 2 vols. 8vo. Rs. 20.

Ram Narain Dass.—A MONOGRAPH OF TWO HUNDRED AND FORTY-EIGHT CASES OF LATERAL LITHOTOMY OPERATIONS performed in the North-Western Provinces during a period of twelve years, by RAM NARAIN DASS. Illustrated by 256 Diagrams of Calculi, showing exterior and section, executed in Lithography. Pp. viii., 94. 8vo., cloth. Rs. 10.

SURVEYING, ETC.

Thuillier.—MANUAL OF SURVEYING FOR INDIA. Detailing the mode of operations on the Trigonometrical, Topographical, and Revenue Surveys of India. By Col. H. L. THUILLIER and Col. SMYTH. Revised and enlarged. Royal 8vo., cloth. Rs. 12.

Mackesy.—UNIVERSAL SCANTLING DIAGRAM. By Col. MACKESY. Price Rs. 4, in case, 4-8. Gives on inspection the correct dimensions for all kinds of timber beams to one tenth of an inch.

Scott.—NOTES ON SURVEYING. For the use of Schools. By J. MIDDLEBROOK SCOTT, C.E. Second Edition. 8vo., stitched. Re. 1.

Kelly.—HANDBOOK OF PRACTICAL SURVEYING FOR INDIA. Especially designed as a Guide and Book of Ready Reference for District Officers, Planters, Municipalities, Courts of Wards and Landholders and for the use of Surveying Classes. Illustrated with Plans, Dirgrams, &c. By F. W. KELLY. Second Edition. Rs. 6.

THEOLOGICAL.

Missions.—STATISTICAL TABLES OF PROTESTANT MISSIONS IN INDIA, BURMA AND CEYLON. Prepared on information collected at the close of 1881, at the request of the Calcutta Missionary Conference, and with the concurrence of the Madras and Bombay Missionary Conferences.

Welland.—EARTHLY AND HEAVENLY TRAINING; Sermons preached to the Calcutta Volunteers. By Rev. JAMES WELLAND. Paper, sewed. Re. 1.

Banerjee.—THE ARYAN WITNESS; or, The Testimony of Aryan Scriptures in Corroboration of Biblical History and the Rudiments of Christian Doctrine, including Dissertations on the Original Home and Early Adventures of Indo-Aryans. By Rev. K. M. BANERJEE. 8vo., paper, Rs. 3-4; cloth, Rs. 4.

Macdonald.—THE VEDIC RELIGION; or, The Creed and Practice of the Indu-Aryans, three thousand years ago. By the Rev. K. S. MACDONALD, M.A. 8vo., pp. 164, sewed. Rs. 1-8.

Slater.—GOD REVEALED. An Outline of Christian Truth, being Sunday Afternoon Lectures delivered to Educated Hindus. By Rev. T. E. SLATER, London Mission, Madras. Re. 1.

Temple.—THE LORD'S PRAYER TRANSLATED INTO THE BOJINGIJIDA, or South Andaman Language. By E. H. MAN. With Preface, Introduction, and Notes by Lieut. R. C. TEMPLE, 21st R. N. B. Fusiliers. Demy 8vo., cloth. Rs. 3-8.

Long.—SCRIPTURE TRUTH IN ORIENTAL DRESS; or, Emblems Explanatory of Biblical Doctrines and Morals. With Parallel or Illustrative References to Proverbs and Proverbial Sayings in the Arabic, Bengali, Canarese, Persian, Russian, Sanskrit, Tamil, Telegu, and Urdu Languages. By Rev. J. LONG. Half bound. Re. 1.

POETRY AND MISCELLANEOUS LITERATURE.

Aliph Cheem.—LAYS OF IND; Poems, Grave and Gay. Seventh Edition, enlarged, with additional Poems and Illustrations by the Author. Cloth, gilt. Rs. 7.

"BASIL ORMOND," and "Christabel's Love:" Metrical Tales of English Life. By ALIPH CHEEM, Author of "Lays of Ind." Rs. 3-8.

THE MEMOIR OF THE LATE JUSTICE ONOOCOOL CHUNDER MOOKERJEE. A Verbatim Reprint. 12mo. Rs. 1-8.

Ghose.—LIBERAL EDUCATION IN INDIA. A Paper read at a Meeting of the Bethune Society on the 21st March, 1878. By NAGENDRA NATH GHOSE, Barrister-at-Law. 8vo., sewed. Re. 1.

INDIAN VIEWS OF ENGLAND: the Effects of Observation of England upon Indian Ideas and Institutions. By NAGENDRA NATH GHOSE. Demy 8vo., stitched. Re. 1.

Long.—THREE THOUSAND BENGALI PROVERBS AND PROVERBIAL SAYINGS ILLUSTRATING NATIVE LIFE AND FEELING AMONG RYOTS AND WOMEN. By Rev. J. LONG. 12mo., paper. As. 8.

Maine.—THE EFEECTS OF OBSERVATION OF INDIA UPON MODERN THOUGHT. By Sir HENRY SUMNER MAINE, K.C.S.I., LL.D., the Rede Lecture, 1875. 8vo. Re. 1.

Dutt.—A VISION OF SUMERU AND OTHER POEMS. By SHOSHEE CHUNDER DUTT. With Portrait of the Author. 8vo., cloth. Rs. 5.

DURGESA NANDINI; or, THE CHIEFTAIN'S DAUGHTER. A Bengali Romance by BANKIM CHANDRA CHATTERJEE. Translated into English Prose by CHARU CHANDRA MOOKERJEE. 8vo., cloth. Rs. 2-8.

STORIES OF THE ITALIAN OPERAS. 8vo., stitched. Rs. 2.

Tawney.—TWO CENTURIES OF BHARTRIHARI. By C. H. TAWNEY, Professor, Presidency College, Calcutta. 12mo., cloth. Rs. 2.

LIFE IN THE MOFUSSIL; or, the Civilian in Lower Bengal. By an EX-CIVILIAN. 2 vols. Rs. 10.

Mitchell.—IN INDIA. Sketches of Indian Life and Travel from Letters and Journals. By Mrs. MURRAY MITCHELL. Sq. 16mo., cloth. Rs. 2-8.

Day.—GOVINDA SAMANTA; or, The History of a Bengal Ryot. By the Rev. LAL BEHARI DAY. Crown 8vc. Rs. 4-4.

Mackay.—SERIOUS REFLECTIONS AND OTHER CONTRIBUTIONS. By late G. ABRIGH-MACKAY, under the *nom de plume* of "Our Political Orphan." Crown 8vo., cloth. Rs. 3.

DURGESHA NANDINI. Itibritta Mulak Upanaya. Shri Bankimchandra Chattopadhyaya Pranita. Roman Akshare Mudrita. Cloth, boards. Price Re. 1, post free Re. 1-3.

This is the first attempt to print a Bengali work in the Roman character. It has been set in type direct from the Bengali original.

SKETCHES IN INDIAN INK. By JOHN SMITH, Jun., Colonel (Retired List). Crown 8vo., cloth. Rs. 5.

Grant.—RURAL LIFE IN BENGAL, illustrative of Anglo-Indian Suburban Life; the Habits of the Rural Classes; the varied Produce of the Soil and Seasons, and the Culture and Manufacture of Indigo. By C. GRANT. Illustrated with 166 Engravings. Royal 8vo., cloth. Rs. 6-8.

ANGLO-INDIAN DOMESTIC LIFE. By the Author of "Rural Life in Bengal." Profusely illustrated. A new Edition abridged and re-written with the Original and other Illustrations selected by the late COLESWORTHY GRANT. (*In preparation.*

DIRECTORIES, EXAMINATION MANUALS, COMMERCIAL TABLES, &c.

THACKER'S BENGAL DIRECTORY, Comprising, amongst other information, Official Directory; Military Directory; Mofussil Directory; Commercial Directory; Trades Directory; Street Directory, Calcutta; Alphabetical List of Residents in Bengal Presidency. Seventeenth Annual Publication. Royal 8vo., cloth, Rs. 14 cash; Rs. 16 credit.

BENGAL OBITUARY; a Compilation of Tablets and Inscriptions; with Directory, Biographical Sketches, and Memoirs. Royal 8vo., cloth. Rs. 2-8.

CIVIL LIST FOR THE LOWER PROVINCES OF BENGAL Published quarterly. Rs. 3.

CALCUTTA UNIVERSITY CALENDAR, containing the Acts incorporating the University and authorizing it to grant Degrees—Bye-Laws—Regulations for all the Examinations, with lists of subjects and Text-Books—The University List of Fellows, Affiliated Institutions, Graduates, &c., and Papers set at the last Examinations in all Degrees. Demy 8vo., cloth. Rs. 5.

Grimley.—RULES FOR THE DEPARTMENTAL EXAMINATIONS OF JUNIOR MEMBERS OF THE COVENANTED CIVIL SERVICE, the Subordinate Executive Service, the Non-Regulation Commission, Police and Opium Departments; including a Selection of Examination Papers, in which are incorporated specimens of questions in Bentham's Theory of Legislation, Political Economy, History of India. For the Examination of Candidates for admission into the Staff Corps, and of Officers in the Political Department. By W. H. GRIMLEY, C.S., Secretary to the Central Examination Committee. 8vo., boards. Rs. 5.

GUIDE TO THE SUBORDINATE CIVIL SERVICE EXAMINATIONS, including a Selection from the Questions set in previous Examinations and the Orders of Government on the subject. By W. H. GRIMLEY. 8vo., boards. Rs. 4.

THE HUNDRED YEARS' ANGLO-CHINESE CALENDAR, from 1776 to 1876. 8vo., stitched. Rs. 7.

Jarrett.—A MANUAL OF REFERENCE TO THE EXAMINATIONS IN ORIENTAL LANGUAGES HELD THROUGHOUT THE BENGAL PRESIDENCY, containing the latest Orders of Government of India, establishing the Tests for Civil and Military Officers, Graded Officers of the Educational Department, and of the Uncovenanted Service, with Specimens of Test Papers for the various Examinations. By H. S. JARRETT, Secretary to the Board of Examiners. Stitched. Rs. 2-8.

Tables.

Merces.—TABLES OF EXCHANGE. A New Edition, showing the Conversion of English Money into Indian Currency and *vice versâ*, calculated for every 16th of a Penny, from 1*s.* 5*d.* to 2*s.* 3*d.* per Rupee. By F. A. D'MERCES. 8vo. Rs. 8.

TABLES OF INTEREST, showing the Interest of any sum from 3 pie to 100,000 rupees, at any rate from 1 to 15 per cent. per annum, reckoning 360 and 365 days to the year. By F. A. D'MERCES. New Edition. Demy 8vo., bound. Rs. 4.

TABLES OF INCOME, WAGES, RENTS, &c., for months of 28, 29, 30, and 31 days; from 4 annas to 10,000 rupees, advancing by 1 rupee to 100 rupees. By F. A. D'MERCES. Fcap. 8vo. Rs. 1-4.

THE HANDY CALCULATOR, showing the cost of goods by number, per dozen, per hundred, per pound, per maund, and at so many seers and chittacks per rupee; also containing Tables of Income, Wages, &c., and Weights. By F. A. D'MERCES. 12mo., pp. 264. Rs. 2.

Spens.—THE INDIAN READY-RECKONER. Containing Tables for ascertaining the value of any number of Articles, &c., from 3 pies to 5 rupees, also Tables of Wages, from 4 annas to 25 rupees, and other useful Tables. By Capt. A. T. SPENS. Rs. 1-12.

Todivala.—MERCHANTS' AND BANKERS' COMPANION; or, Tables for calculating Interest and Commission, Freight and Exchanges between India, England, and China, in Rupees, Sterling, and Dollars. With Tables to show the Rates of Exchange and Prices of Bullion. By MANEKJI KAVASJI TODIVALA. Royal 8vo. Rs. 12.

Trott—RAILWAY WEIGHT CALCULATOR; Tables for converting tons, cwts., lbs. into maunds and seers; maunds and seers into tons, cwts.. and lbs.; kilogrammes into maunds and seers: kilogrammes into tons, cwts., and lbs. Computed and compiled by JAMES HARINGTON TROTT. 4to. As. 8; post-free 9.

THE LANGUAGES OF INDIA.

Kayathi.

(The language lately ordered to be adopted in the Courts of Behar.)

A HANDBOOK TO THE KAYATHI CHARACTER. By G. A. GRIERSON, Esq., C.S., late Sub-Divisional Officer of Madhubani Tirhoot. Royal 4to., cloth. Rs. 7-8.

The works contains thirty plates of facsimile handwriting in various styles of writing, with transliteration and translation to each.

Hindustani.

The Popular Dictionary: English-Hindustani, Hindustani-English. Hindustani Proverbs, Prefixes, and Affixes in Urdu. 20,000 words. Rs. 1-4; post-free 1-6.

ARAISH-I-MAHFIL. Printed in Urdu. For the use of the Junior Members of Her Majesty's Indian Civil and Military Services. Fourth Edition. Revised and corrected by W. NASSAU LEES, LL.D. 8vo., paper. Rs. 6.

FISANAH-I-AJAIB OF MIRZA ALI BEG SURER. New and Revised Edition. For Examination for Degrees of Honour in Urdu. 8vo. Rs. 4.

Lower Standard.—HINDUSTANI COLLECTIONS FOR THE LOWER STANDARD EXAMINATION: being Selections from the Bagh-o-Bahar and History of India. 8vo. Rs. 2-8.

Higher Standard.—HINDUSTANI SELECTIONS FOR THE HIGHER STANDARD; being Selections from the Bagh-o-Bahar and Premsagur. 8vo., stitched. Rs. 4-8.

IKHWAN-US-SAFA, IN URDU. Fourth Edition, revised and corrected. For Examination, High Proficiency in Urdu. 8vo. Rs. 3.

NUZR-I-BE-NUZEER, IN URDU. Third Edition, revised and corrected. For Examination, High Proficiency in Urdu. 8vo. Rs. 3.

NAZR-I-BI-NAZIR. Translated from Urdu by C. W. BOWDLER BELL. 8vo., stitched. Rs. 4-8.

NAZR-I-BI-NAZIR; or, The Incomparable Prose of Meer Hussen. Literally translated by Capt. M. H. COURT. 8vo., stitched. Rs. 5-8.

Stapley.—GRADUATED TRANSLATION EXERCISES. Part II—English-Urdu, Urdu-English. With Rules, and Remarks for the guidance and assistance of Native Teachers and Students. By L. A. STAPLEY. 8vo., paper. Ans. 6.

KULLIYAT-I-SAUDA OF MIRZA RAFI-US-SAUDA, Extracts from, in Urdu. For Examination for Degrees of Honour in Urdu. 4to. Rs. 4.

MUSNAWIYAT OF MIRZA RAFI-US-SAUDA. For Candidates for the High Proficiency Examination in Urdu. Sewed. As. 12.

Court.—SELECTIONS FROM THE KULLIYAT-I-SAUDA. The parts appointed for the High Proficiency Examination. Literally translated by Capt. M. H. COURT. 8vo., paper. Rs. 3.

Yates.—INTRODUCTION TO THE HINDUSTANI LANGUAGE. Eighth Edition, improved. By Rev. W. YATES. In Roman Character, Rs. 5 In Persian Character, Third Edition, Rs. 6.

URDU ENTRANCE COURSE. Selections appointed by the Calcutta University. Poetry, As. 12; Prose, As. 10.

THE PEOPLE'S DICTIONARY IN ENGLISH AND ROMAN URDU. An Etymological, Idiomatic, and Illustrated Dictionary. 8vo., boards. As. 8; post free 9.

THE ROYAL SCHOOL DICTIONARY IN ENGLISH AND ROMAN URDU. Etymological, Pronouncing (English), and Idiomatic. Crown 8vo., boards. Re. 1; post free, 1-2.

Jarrett.—TRANSLATION OF THE HINDU PERIOD OF MARSHMAN'S HISTORY for the Lower Standard Examination. By Major H. S. JARRETT. 8vo. Rs. 1-4.

Bell.—THE RAJNITI OF LALLU LALA. Translated from the Braj Bhasha by C. W. BOWDLER BELL. 8vo., stitched. Rs. 6.

BYTAL PACHEESEE: or, The Twenty-five Tales of a Demon. Reprinted from the New Edition of Eshwar Chandra Vidyasagar in Hindi. Second Edition. Rs. 3.

Thompson.—A DICTIONARY IN HINDI AND ENGLISH. Compiled from approved Authorities. Second Edition. By J. T. THOMPSON. Royal 8vo., paper. Rs. 10.

Sanscrit.

Tawney.—UTTARA RAMA CHARITTA, a Sanscrit Drama by Bhavabhute. Translated into English Prose by C. H. TAWNEY, Professor, Presidency College, Calcutta. Sewed. Rs. 1-8.

THE MALAVIKAGNIMITRA: a Sanscrit Play by Kalidasa. Literally translated into English Prose by C. H. TAWNEY, Professor, Presidency College, Calcutta. 12mo., cloth. Rs. 2.

TWO CENTURIES OF BHARTRIHARI. By C. H. TAWNEY, Professor, Presidency College, Calcutta. 12mo., cloth. Rs. 2.

Banerjee.—BHUTTI KAVYA: Poem on the Actions of Rama. The first five Books, with Notes and Explanations. By Rev. K. M. BANERJEE. 8vo., paper. Rs. 2.

KUMAR SAMBHAVA OF KALIDASA. With Notes and Explanations in English. By Rev. K. M. BANERJEE. Third Edition, revised. 8vo., paper. Rs. 2-8.

RAGHUVANSA BY KALIDASA. With Notes and Grammatical Explanations. By Rev. K. M. BANERJEE. 8vo. paper. No. 1 (Cantos 1—3), Rs. 2. No. 2 (Cantos 4—9), Rs. 3. The two parts together, Rs. 4.

Ram Jasan.—A SANSCRIT AND ENGLISH DICTIONARY; being an abridgment of Professor Wilson's Dictionary. With an Appendix explaining the use of Affixes in Sanscrit. By PANDIT RAM JASAN. Royal 8vo., cloth. Rs. 10.

WILSON'S DICTIONARY IN SANSCRIT AND ENGLISH; translated, amended, and enlarged from an original Compilation prepared by learned Natives for the College of Fort William. Third Edition. By JAGUN MOHANA TARKALANKA and KHETTER MOHANA MUKERJEE. 4to., bound. Rs. 16.

Persian.

A'ARAS-I-BOZORGAN; being an Obituary of Pious and Learned Moslims, from the beginning of Islam to the middle of the 12th Century of the Hijirah. Edited in Persian by Col. W. N. LEES and KABIR ALDIN AHMED. 8vo., paper. Rs. 2-8.

GULISTAN OF SADY. Edited in Persian, with Punctuation and the necessary Vowel-marks. Third Edition, revised and corrected. By Col. W. N. LEES. For Examination, High Proficiency in Persian. 8vo. Rs. 5.

IQD-I-GUL; being a Selection from the Gullistan and Anwari Sohaili. For Examination, Higher Standard. Rs. 3.

IQD-I-MANZUM; being a Selection from the Bostan of Sadi. Edited by KABIR ALDIN AHMED. Second Edition. For Examination, Higher Standard. 8vo. Rs. 2.

SELECTIONS FROM THE SHAH NAMAH. For the Degree of Honour Examination in Persian. For Officers in the Military and Civil Services. By Major H. S. JARRETT. Published by Authority.

Lee's Persian Series:—

I.—ZAWABIT-I-FARSI; or, A Persian Grammar in Oordoo. Prepared by ABD-UL-HUQQ. Seventh Edition. 12mo. As. 5.

IV.—IQD-I-LAALI; or, A Persian Poetical Reader, No. 1, containing the Pand-Nama or Karima and the Tarjiband, called Mamoquiman. Second Edition. 12mo. As. 4.

V.—IQD-I-MARJAN; or, Persian Poetical Reader, No. 2, containing the Pand-Nama of Farid Aldin Attar and the Quiccah-i-Yoosof; being an extract from Jami's Poem Yoosof and Zalaikha. Second Edition. 8vo. As. 8.

MODERN PERSIAN PHRASES. By an Officer of the Hyderabab Contigent. Revised by AKA MIRZA ZENUL ABIDEEN SHEERAZEE. 8vo., cloth. Rs. 5.

PERSIAN SELECTIONS FOR FIRST ART COURSE, Rs. 1-4; For B.A. Course, Rs. 2-8.

Arabic.

ALIF LAILA (THE): or, Book of the Thousand Nights and one Night. Edited by Sir W. H. MACNAGHTEN. 4 vols. royal 8vo., (all remaining copies soiled). Rs. 25.

ARABIC SELECTIONS FOR THE ENTRANCE EXAMINATION entitled Sullam-ul-Adah. 8vo., cloth. Rs. 3-8.

ARABIC SELECTIONS FOR THE FIRST ARTS EXAMINATION, 8vo., cloth. Rs. 2-8.

Hayes.—SYNOPSIS OF THE GRAMMAR OF MODERN ARABIC. With a Vocabulary and Dialogues. By Capt. F. F. C. HAYES. Fcap. 8vo., paper. Rs. 2.

KALUBI, IN ARABIC. Edited by Col. W. N. LEES and MAULAVI ABD-UL-HUQQ. 8vo. Rs. 2.

KASHAF; being the Commentary on the Koran. By ABL-KASIM MUHMAD. 2 vols. 4to. Rs. 25.

Lee's Arabic Series:—

I.—ARABIC INFLECTION. 12mo. As. 4.

II.—ARABIC SYNTAX. 12mo. As. 4.

III.—TUHFAT-UL-TALIBIN. 12mo. As. 4.

Pukshto.

Bellew.—A DICTIONARY OF THE PUKHTO OR PUKSHTO LANGUAGE. In which the words are traced to their sources in the Indian and Persian Languages. By H. W. BELLEW. 4to., cloth. Rs. 21.

A GRAMMAR OF THE PUKHTO OR PUKSHTO LANGUAGE on a new and improved system, combining brevity with practical utility, and including Exercises and Dialogues intended to facilitate the acquisition of the Colloquial. By H. W. BELLEW. 4to., cloth. Rs. 10-8.

Vaughan.—A GRAMMAR OF THE PUKSHTO LANGUAGE as spoken in the Trans-Territories under British Rule. By Capt. J. L. VAUGHAN. 8vo., paper. Rs. 4.

Bengali.

STAPLEY'S GRADUATED TRANSLATION EXERCISES adapted to the Bengali Language. 8vo. As. 8.

THE LITERATURE OF BENGAL, from the earliest times to the present day. By ARCYDAE. With Extracts. 12mo., stitched Re. 1.

Mendes. — ABRIDGMENT OF JOHNSON'S DICTIONARY, English and Bengali. Second Edition. By JOHN MENDES. 8vo., cloth. Rs. 5.

YATES'S INTRODUCTION TO THE BENGALI LANGUAGE, Edited by J. WENGER. Vol. I.—Grammar and Reader, 8vo., cloth. Rs. 5; Vol. II.—Selection from Bengali Literature. 8vo. Rs. 5.

Glossaries.

Carnegy. — KACHARI TECHNICALITIES; or, A Glossary of Terms—Rural, Official, and General—in daily use in the Courts of Law, and in illustration of the Tenures, Customs, Arts, and Manufactures of Hindustan. By P. CARNEGY. Demy 8vo., cloth. Rs. 12.

Durga Prasad.—A GUIDE TO LEGAL TRANSLATION; or, A Collection of Words and Phrases used in the Translation of Legal Papers from Urdu to English. By DURGA PRASAD. 8vo., cloth. Rs. 5.

Rajendra Lala Mittra.—A SCHEME FOR THE RENDERING of European Scientific Terms into the Vernacular of India. By RAJENDRA LALA MITTRA. Demy 8vo., stitched. As. 8.

Robinson.—DICTIONARY OF LAW AND OTHER TERMS commonly employed in the Courts of Bengal, including many Commercial Words and Idiomatic Phrases, in English and Bengali. By JOHN ROBINSON. 8vo. Rs. 6.

EDUCATIONAL WORKS.

Including Macmillan & Co's English Series of "Text books for Indian Schools" (English prices at the Exchange of the day).

ENGLISH LANGUAGE.

A MORAL READING BOOK FOR INDIAN SCHOOLS. Selected from British and Oriental Sources. By ROPER LETHBRIDGE. Foolscap 8vo. As. 14.

MILTON'S ODE ON THE MORNING OF CHRIST'S NATIVITY With Notes (Todd's). Fcap. 8vo., limp. As. 4.

A GUIDE TO ENGLISH PARSING. By H. P. JACOB. Fcap. 8vo. As. 4.

LAMB'S TALES FROM SHAKESPEARE. Thacker, Spink & Co.'s School Edition. Fcap., cloth. As. 12.

SELECTIONS PROM MODERN ENGLISH LITERATURE. By E. LETHBRIDGE. 8vo., cloth. Rs. 2.

FIRST ARTS COURSE FOR 1881. Authorized by the Calcutta University. 8vo., cloth. Rs. 5.

ENGLISH CLASSICS FOR INDIAN SCHOOLS: Goldsmith's Deserted Village. Edited by G. W. FORREST and STORR. As. 12.

EASY SELECTIONS from Modern English Literature. By E. LETHBRIDGE. 1*s*. 6*d*.

A SERIES OF SIX ENGLISH READING-BOOKS. For Indian Children. By PEARY CHURN SIRCAR. Revised by E. LETHBRIDGE. First 5*d*.; Second 6*d*.; Third, 8*d*.; Fourth, 1*s*.; Fifth, 1*s*. 2*d*.; Sixth, 1*s*. 3*d*.

HINTS ON THE STUDY OF ENGLISH. By F. J. ROWE and W. T. WEBB. Demy 8vo., cloth. Rs. 2-8.

A COMPANION READER to "Hints on the Study of English." Demy 8vo., paper. Rs. 1-4.

THE STUDENT'S COMPANION TO ENGLISH GRAMMAR. Compiled from University Examination Papers. By J. TANNOCK, Nizamut College, Moorshedabad. 8vo., stitched. As. 8; Supplement, 1876.

A KEY TO TANNOCK'S STUDENT'S COMPANION TO ENGLISH. By G. A. LORRIMER. 8vo., sewed. As. 8.

ENGLISH PEOPLE AND THEIR LANGUAGE. Translated from the German of Loth by C. H. TAWNEY. Stitched. As. 8.

A PRIMER OF ENGLISH GRAMMAR. For Indian Students. By the Rev. R. MORRIS, M.A., LL.D. 1*s.*

A SHORT HISTORY OF ENGLISH LANGUAGE. By THOMAS EDWARDS, F.E. Fcap. 8vo. Rs. 1-4.

MacMillan—QUESTIONS AND ANSWERS ON ELEMENTARY LOGIC. By M. MACMILLAN, B.A., Professor of Logic and Moral Philosophy at Elphinstone College, Bombay. Fcap. 8vo. Re. 1; post-free 1-1.

"All the Answers follow closely the lines laid down in Fowler's Deductive Logic."

GEOGRAPHY.

A GEOGRAPHICAL READER AND COMPANION TO THE ATLAS. By C. B. CLARKE, M.A. 2*s.*

PHYSICAL GEOGRAPHY. For Indian Students. By H. F. BLANFORD, Associate of the Royal School of Mines, Director-General of the Indian Meteorological Department. 2*s.* 6*d.*

THE INDIAN ATLAS OF THE WORLD. By JOHN BARTHOLOMEW. Coloured. 16 Maps. As. 6, post free 6½.

INDIAN SERIES OF MAPS. Imperial 4to., prepared from the best authorities, and printed in colours. Price one anna each; or the set, post-free, for one rupee. List of Maps.—Hemispheres; Europe; Asia; Africa; North America; South America; British Isles; England and Wales; Persia, Afghanistan, and Beloochistan; India; North West Provinces; Punjab, &c.; Chinese Empire and Japan; Palestine; United States.

HISTORY.

Macleod—A CONCISE EPITOME OF TAYLOR'S ANCIENT HISTORY. By HERBERT G. W. MACLEOD. Re. 1-8; post-free 1-10.

THE WORLD'S HISTORY. Compiled under the direction of E. LETHBRIDGE, M.A. Second Edition. 1*s.*

AN EASY INTRODUCTION TO THE HISTORY OF INDIA Second Edition. By E. LETHBRIDGE, M.A. 1*s.* 6*d.*

A HISTORY OF ENGLAND. Second Edition. Compiled under the direction of E. LETHBRIDGE. M.A., 1*s.* 6*d.*

AN EASY INTRODUCTION TO THE HISTORY AND GEOGRAPHY OF BENGAL. Second Edition. By E. LETHBRIDGE, M.A. 1*s.* 6*d.*

MATHEMATICS.

EUCLID'S ELEMENTS OF GEOMETRY. Part I., containing the First Four Books, with Notes, &c. By P. GHOSH. Sewed, Rs. 1-4; cloth, Rs. 1-8.

ARITHMETIC WITH ANSWERS. For Indian Students. By BARNARD SMITH, M.A. 2*s.*

This Edition of Barnard Smith's Arithmetic contains the Indian Tables of Weights, Measures, and Currency, as well as the English Tables; and Examples fully worked out, and very numerous Questions for practice, based on the Tables, have been inserted under each rule.

A CHAPTER OF DECIMALS. By J. M. ALDIS, M.A. 16mo Stitched. Re. 1.

ARITHMETIC FOR BEGINNERS, designed for the use of Junior Classes, to which are appended Exercises in Mental Arithmetic. By CHARLES D'CRUX, Bengal Academy. 16mo. As. 6.

SOLUTIONS OF EXERCISES FROM TODHUNTER'S MENSURATION. Including Surveying, and an Appendix. By P. GHOSH 12mo. Rs. 1-8

ELEMENTARY TREATISE ON MECHANICS. By F. W. STEPHENSON. Stitched, 8vo. Rs. 2-8.

ALGEBRA. For Indian Students. By I. TODHUNTER, M.A., F.R.S. 2*s.* 6*d.*

THE FIRST FOUR BOOKS OF EUCLID. With Notes, Appendix, and Exercises. By I. TODHUNTER, F.R.S. 2*s.*

ELEMENTARY MENSURATION AND LAND SURVEYING. For Indian Students. By I. TODHUNTER, M.A., F.R.S. 2*s.*

This Edition of Todhunter's Mensuration has been prepared specially for Entrance Candidates, and contains an Appendix on Practical Surveying in India.

NOTES ON PRACTICAL GEOMETRY AND THE CONSTRUCTION OF SCALES. By J. M. SCOTT. Second Edition. 8vo., stitched. Re. 1.

THE ELEMENTS OF ALGEBRA, by JAMES WOOD; remodelled, simplified, and adapted for general use in schools, with numerous Exercises and Examination Papers. Part I. For Entrance Candidates. By P. GHOSH. 12mo., cloth. Rs. 1-8.

MISCELLANEOUS ARITHMETICAL QUESTIONS selected from various sources. With neat and concise Solutions. By EDWARD FELL. 12mo. Rs. 2.

ELEMENTARY TREATISE ON PLANE TRIGONOMETRY. With Examination Papers and Examples. Designed for the use of the Government Colleges of India, and Schools affiliated to the University of Calcutta. By E. WILMOT. Pp. 76. 8vo., cloth. Rs. 2-12.

ALGEBRAICAL EXERCISES WITH SOLUTIONS. For Students preparing for the Entrance Examination of the Calcutta University. By SARAT CHANDRA MUKHOPADHYAY, M.A. 12mo., stiff cover. Rs. 1-4.

ELEMENTARY DYNAMICS, with numerous Examples. By W. G. WILLSON. Second Edition, by J. WILLSON. Crown 8vo. Rs. 3-8.

BENGALI.

BENGALI ENTRANCE COURSE. Entrance Examination. Rs. 1-6.

THE WORLD'S HISTORY. Compiled under the direction of E. LETHBRIDGE, M.A. Translated into Bengali. Demy 8vo. As. 8.

AN EASY INTRODUCTION TO THE HISTORY AND GEOGRAPHY OF BENGAL. For the Junior Classes in Schools. By E. LETHBRIDGE, M.A. Translated into Bengali. Demy 8vo, Re. 1.

A PRIMER OF CHEMISTRY. By H. E. ROSCOE. Translated into Bengali. Demy 8vo. As. 7.

STAPLEY'S GRADUATED TRANSLATION EXERCISES, adapted to the Bengali Language. As. 8.

ELEMENTS OF EUCLID. First two Books With Notes and Exercises. Translated into Bengali. Demy 8vo. As. 6.

MENSURATION FOR BEGINNERS. With numerous Examples By I. TODHUNTER, M.A., F.R.S. Translated into Bengali. Demy 8vo. As. 12.

EXTRACTED CHAPTERS FROM THE RUDIMENTS OF PHYSICAL GEOGRAPHY FOR INDIAN SCHOOLS. By H. F. BLANFORD. Translated into Bengali by BABOO PURNA CHANDRA DATTA for the Author. As. 10.

AN EASY INTRODUCTION TO THE HISTORY OF INDIA. By E. LETHBRIDGE, M.A. Translated by BABOO RADHIKA PRASANNA MUKERJEE. Demy 8vo. Re. 1.

METAPHYSICS.

THE ELEMENTS OF THE PSYCHOLOGY OF COGNITION. By R. JARDINE. Crown 8vo. 6*s.* 6*d.*

ANALYSIS OF SIR WILLIAM HAMILTON'S LECTURES ON METAPHYSICS., with Dissertations and Copious Notes. By W. CAREY FINK. Author of "Analysis of Dr. Thos. Reid's Inquiry into the Human Mind." Rs. 2.

ANALYSIS OF REID'S INQUIRY INTO THE HUMAN MIND. By W. C. FINK. With Notes. Second Edition. 8vo., stitched. Rs. 1-12.

LECTURES ON METAPHYSICS. By Sir WILLIAM HAMILTON, Bart., 2 vols. 8vo., cloth. Rs. 9.

AN INQUIRY INTO THE HUMAN MIND ON THE PRINCIPLES OF COMMON SENSE. By THOMAS REID, D.D. 8vo., cloth. Rs. 1-4.

LATIN.

SELECTA POETICA. The Selection adopted by the Calcutta University. Part I., Selections from Ovid. 18mo., cloth. Re. 1. Translation. Re. 1.

SELECTA POETICA. The Selections adopted by the Calcutta University. Part II., Selections from Catullus, Tibullus, Propertius, Martial, Aulus, Persius, Juvenal. 18mo., cloth. Re. 1. Translation. Re. 1.

ARNOLD'S ECLOGÆ OVIDIANÆ. A Translation. Fcap. 8vo., sewed. Re. 1.

INDIAN LAW PUBLICATIONS.

CIVIL PROCEDURE.

Code of Civil Procedure (The); being Act X. of 1877. With Notes and Appendix, containing Charters of all the High Courts, Privy Council Appeals, Orders, Rules, the Limitation and Specific Relief Acts; with Copious Index. By L. P. DELVES BROUGHTON, assisted by W. F. AGNEW and G. S. HENDERSON. Royal 8vo. Reduced to Rs. 18. Appendix, Act XII. of 1879. Rs. 1.

Civil Procedure Code; Act XIV. of 1882. With copious Index. Royal 8vo. Rs. . Interleaved.

COMMERCIAL.

The Contract Act (IX. of 1872). With an Introduction and Explanatory Notes. By the Hon. H. S. CUNNINGHAM, M.A., and H. H. SHEPHARD, M.A. Third Edition. Demy 8vo., cloth. Rs. 14.

The Indian Contract Act (IX. of 1872). With a Commentary, critical, explanatory, and illustrative. By C. C. MACRAE. 8vo., cloth. Rs. 10.

The Indian Contract Act (IX. of 1872), and the Specific Relief Act (I. of 1877). With a full Commentary. By D. SUTHERLAND. Royal 8vo., cloth. Rs. 10.

The Indian Companies' Act, 1866 (Act X. of 1866). With an Appendix and Notes. By WHITLEY STOKES. Royal 8vo. Rs. 8.

Law of Indian Railways and Common Carriers. A Commentary upon the Indian Railway Act of 1879, the Carriers' Act of 1865, and the Act known as Lord Campbell's Act (XIII. of 1855). Together with such Sections and Quotations from the Merchant Shipping Act, the Contract Act, the Railway and Canal Traffic Act, Civil Procedure Code, Court Fees Act, and the Statute of Limitations, as apply to Carriers by Land and Water. By W. G. MACPHERSON. 8vo., cloth. Rs. 12.

The Sea-Customs Law, 1878, and Tariff Act; with Notes and Appendices. By W. H. GRIMLEY, Esq., B.A., LL.B., C.S. Demy 8vo., cloth, Rs. 7-8; Interleaved, Rs. 8-8.

Commercial—*continued.*

The Indian Registration Act (III. of 1877). With Notes (fourth edition of the Notes). By CARR-STEPHEN, Barrister-at-Law. Demy 8vo., cloth. Rs. 5.

A Treatise on the Law of Bills of Lading. Comprising the various Incidents attaching to the Bill of Lading, the Legal Effects of the Clauses and Stipulations, the Rights and Liabilities of Consignors, Consignees, Indorsees, and Vendors, under the Bill of Lading. With an Appendix, containing Forms, &c. By EUGENE LEGGETT, Solicitor. 8vo., cloth. Rs. 10; post free 10-8.

Negotiable Instruments Act (1881). With Explanatory Commentary, comparison of the Indian and English Laws and the Continental Codes, &c. By M. B. CHALMERS, author of "Digest of the Law of Bills of Exchange, Cheques," &c. 8vo. Rs. 7.

Indian Case-Law on Torts. By R. D. ALEXANDER, B.C.S., Offg. Judge, Small Cause Court, Allahabad. Crown 8vo., cloth. Rs. 4.

Coolie Act. The Inland Emigration Act, 1882. 8vo. Interleaved. R. 1.

Manual of Indian Criminal Law. Being the Penal Code, Criminal Procedure Code, Evidence Act, Whipping Act, Police Act, &c. With Annotations and all applicable Rulings, arranged under the appropriate sections. By H. A. D. PHILLIPS, C.S.

CRIMINAL LAW.

The Indian Penal Code Act (XLV. of 1860), and other Laws and Acts of Parliament relating to the Criminal Courts of India. With Notes containing the Rulings of the Nizamut Adawlut on points of Procedure and Decisions of the High Court of Calcutta. By J. O'KINEALY. Second Edition. Royal 8vo. Rs. 14.

The Indian Magistrate's Guide: an Introduction to the Judicial Duties of Magistrates and Justices of the Peace in India. By Sir BENSON MAXWELL. Edited for India by L. P. DELVES BROUGHTON, B.A., Barrister-at-Law. 8vo., cloth. Rs. 12.

Compendium of Criminal Rulings. With a copious Index, a Table of Cases and Notes. By KEDAR NATH GHOSH, B.L. Part I. Reduced in price. Sewed. Rs. 2; in cloth, 2-4, and postage As. 4. Part II. 1880. Re. 1; post-free 1-1.

The Penal Code Act (XLV. of 1860) (Fifth Edition), as amended by later Enactments. With Rulings of all the High Courts and Chief Courts in India. By FENDALL CURRIE. 8vo., cloth. Rs. 8.

Pocket Penal, Criminal Procedure, and Police Codes; also the Whipping Act, and the Railway Servants Act: being Acts XLV. of 1860 (with Amendments), X. of 1882, V. of 1861, VI. of 1864, and XXXI. of 1867. Rs. 4

Criminal Law—*continued.*

Code of Criminal Procedure: being Act X. of 1882. With Table of Contents and Index. Rs. 4. Interleaved, Rs. 5.

The Code of Criminal Procedure (Act X. of 1882), together with Rulings, Circular Orders, Notifications, &c., of all the High Courts in India, and Notifications and Orders of the Government of India, and the Local Governments. Edited, with Copious Notes and Full Index, by W. F. AGNEW, Barrister-at-Law, Author of "A Treatise on the Law of Trusts in India," and GILBERT S. HENDERSON, M.A., Barrister-at-Law, Author of "A Treatise on the Law of Testamentary and Intestate Succession in India."

EVIDENCE.

The Indian Evidence Act (I. of 1872). With an Introduction on the Principles of Judicial Evidence. By Sir JAMES FITZJAMES STEPHEN. 8vo. Rs. 9.

The Law of Evidence (Act I. of 1872, with Amendments) With a Commentary, Notes, &c. Third Edition. By C. D. FIELD. Demy 8vo., cloth. Rs. 18.

LAND.

Law of Mortgage in Bengal and the North-West Provinces. By A. G. MACPHERSON, Judge of the High Court of Judicature. Sixth Edition. Demy 8vo. Rs. 10.

Revenue Sale Law of Lower Bengal, comprising Act XI. of 1859: Bengal Act VII. of 1868; Bengal Act VII. of 1880. (The Public Demands Recovery Act) and the Unrepealed Regulations and the Rules of the Board of Revenue on the subject. With Notes. Edited by WILLIAM E. H. FORSYTH. Demy 8vo., cloth. Rs. 5; post free Rs. 5-4.

The Rent Law of Bengal, L. P. (Act VIII. 1869, B.C.). with all important Rulings and Notes. By JOGINDRA CHANDRA MAULIK. 8vo., cloth. Rs. 4.

The Law of Landlord and Tenant as administered in Bengal. By H. BELL, C.S., Barrister-at-Law. Second Edition. Cloth, large 8vo. Rs. 4.

The Land Acquisition Act, being Act X of 1870. By H. BEVERLEY, Esq., B.C.S. With Notes. Demy 8vo., cloth. Rs. 5.

LEGISLATIVE ACTS.

Bengal Acts: the Legislative Acts (Unrepealed) of the Lieutenant-Governor of Bengal in Council. From 1862 to the present. Annotated and edited with Index, by FRED. CLARKE, Esq., Barrister-at-Law. Rs. 12.

Legislative Acts published annually in continuation of Mr. Theobald's Edition. Royal 8vo., cloth—1872, Rs. 8; 1873, 1874, 1875, Rs. 5 each; 1876, Rs. 6; 1877, Rs. 10; 1878, Rs. 5; 1879, Rs. 5; 1880, Rs. 4. 1881, Rs. 8.

Chronological Tables of Acts of Parliament. Regulations of the Bengal Code, Acts of the Governor-General of India in Council, and Acts of the Lieutenant-Governor of Bengal in Council, with Remarks. By H. S. CUNNINGHAM, Esq., one of the Judges of the High Court of Judicature, Bengal.

Statutes, Regulations, and Acts in force in Bengal. Arranged chronologically and also according to the nature of their Contents. By the Hon. H. S. CUNNINGHAM, M.A. *(In the Press.)*

Statutes, Regulations, and Acts in force in Bengal. Tables of Acts of Parliament, Regulations of the Bengal Code, Acts of the Governor-General of India in Council, and Acts of the Lieutenant-Governor of Bengal in Council, arranged Chronologically, and also according to the nature of their contents. By H. S. CUNNINGHAM, Esq., one of the Judges of the High Court of Judicature, Bengal. *(In the Press.)*

ORIENTAL LAW.

The Law relating to the Hindu Widow. By TRAILOKYANATH MITTRA, M.A., D.L. Royal 8vo., cloth. Rs. 10.

The Principles of the Hindu Law of Inheritance; together with I—A description, and an inquiry into the origin, of the *Sraddha* Ceremonies, II—An account of the historical development of the Law of Succession, from the Vedic period to the present time; and III—A Digest of the Text-Law and Case-Law bearing on the subject of Inheritance. The Tagore Law Lectures, 1880. By RAJCOOMAR SARVADHICARI, B.L., Tagore Law Professor, Law Lecturer and Professor of Sanskrit, Canning College, Lucknow. Royal 8vo., cloth. Rs. 16.

The Personal Law of the Mahomedans (according to all the Schools); together with a comparative sketch of the Law of Inheritance among the Sunnis and the Shiahs. By SYED AMEER ALI, Moulvi, M.A., LL.B. 8vo. Rs. 10-12.

Hindu Law.—A Prospectus of the Scientific Study of Hindu Law. By I. H. NELSON, M.A. Rs. 6-8.

REPORTS.

Indian Law Reports, published under Authority. The Reports appear in Monthly Parts, at Calcutta, Madras, Bombay, and Allahabad, and comprise four Series.—The Cases heard by the Privy Council on appeal from each High Court are reported in the Series for that High Court. Annual Subscription, Rs. 47-8.

Judgments of the Privy Council on Appeals from India. 1831 to 1877. By D. SUTHERLAND. In paper cover, Rs. 16. Vol. II., 1868 to 1877, Rs. 20. Vol. III., Rs. 20. The three vols. together, Rs. 40.

Moore's Indian Appeals, a Reprint, 14 vols. half-bound, Rs. 210.

The Digest of Indian Law Reports: a Compendium of the Rulings of the High Court of Calcutta from 1862, and of the Privy Council from 1831 to 1876. By D. SUTHERLAND, Esq., Barrister-at-Law. Impl. 8vo. Rs. 20.

The Digest of Indian Law Reports. Vol. II., 1876 to 1881. Rs. 16.

The Indian Digest: a complete Index to the Reported Cases of the High Courts established in India. Second Edition. By H. COWELL, Barrister-at-Law. Royal 8vo., cloth. Reduced to Rs. 12.

Digest of Cases reported in the Indian Law Reports: Calcutta Series, Vol. I. to IV.; Bombay Series, Vol. I. to III.; Allahabad Series, Vols. I. and II.; Madras Series, Vols. I. and II.; and L. R. Indian Appeals, Vols. I. to VI. Compiled by REGINALD M. A. BRANSON. Royal 8vo., cloth. Rs. 21.

Digest of Cases reported in the Bombay High Court Reports, Vols. I. to XII.; and the Indian Law Reports, Bombay Series, Vols. I. and II. Compiled by REGINALD M. A. BRANSON. Royal 8vo., cloth. Rs. 16.

A Digest of all the Cases relating to India decided by the Privy Council and the other Courts in England. By REGINALD M. BRANSON. Royal 8vo., cloth. Rs. 20.

SOCIAL.

The Indian Succession Act, 1865 (Act X. of 1865). With a Commentary, and the Parsee Succession Act, 1865, Acts XII. and XIII. of 1855, and the Acts relating to the Administrator-General. With Notes. By WHITLEY STOKES. 8vo. Rs. 10.

The Probate and Administration Act; being Act V. of 1881. With Notes. By W. E. H. FORSYTH, Esq. Edited with Index by F. J. COLLINSON, Esq. Demy 8vo., cloth. Rs. 5.

The Law of Divorce in India; being the Indian Divorce Act. With Notes and Forms of Pleadings. By C. C. MACRAE. Cloth. Rs. 10.

The Indian Registration Act (III. of 1877). With Note. By CARR-STEPHEN. Fourth Edition. 8vo. Rs. 5.

The Law of Intestate and Testamentary Succession in India; including the Indian Succession Act, 1865, with a commentary; Parsee Succession Act, 1865, Hindu Wills Act, 1870, Probate and Administration Act, 1881, District Delegates Act, 1881, Acts XII. and XIII. of 1855, Regimental Debts Act, 1863 (26 and 27 Vict. c. 57), Acts relating to the Administrator-General, the Certificate Act, 1860, and the Oudh Estates Act, 1869. With Notes and Cross References. By GILBERT S. HENDERSON, M.A. Royal 8vo., cloth. Rs. 16.

SPECIFIC RELIEF.

The Law of Specific Relief in India.—Being a Commentary on Act I. of 1877. By C. COLLETT. Rs. 9.

Specific Relief Act.—Commentaries on the (Act No. I. of 1877), containing ample Annotations founded on eminent English and American works on Equity Jurisprudence, and on the English Law Reports, together with Notes on the Indian Decisions, and with Table of Cases and copious Index. By VENKATRAO RAMCHANDRA. Rs. 10-8.

STUDENTS.

The Indian Law Examination Manual. Second Edition. Revised with all the new Laws. By FENDALL CURRIE. 8vo. Rs. 5.

Glossary of Medical and Medico-Legal Terms; including those most frequently met with in the Law Courts. Compiled by R. F. HUTCHINSON, M.D. Foolscap 8vo. Rs. 3. Post free 3-2.

TAGORE LAW LECTURES.

The Hindu Law; being a Treatise on the Law administered exclusively to Hindus by the British Courts in India. (1870.) By HERBERT COWELL. Royal 8vo., cloth. Rs. 12.

The Hindu Law; being a Treatise on the Law administered exclusively to Hindus by the British Courts in India. (1871.) By HERBERT COWELL. Royal 8vo., cloth. Rs. 8.

History and Constitution of the Courts and Legislative Authorities. (1872.) By HERBERT COWELL. Royal 8vo., cloth. Rs. 8.

Mahomedan Law; a Digest of the Laws applicable to the Sunnis of India. (1873.) By SHAMA CHURN SIRCAR. Royal 8vo., cloth. Rs. 9.

Mahomedan Law; being a digest of the Sunni Code in part, and of the Imamyah Code. (1874.) By SHAMA CHURN SIRCAR. Royal 8vo., cloth. Rs. 9.

The Law relating to the Land Tenures of Lower Bengal. (1875.) By ARTHUR PHILLIPS. Royal 8vo., cloth. Rs. 10.

The Law relating to Mortgage in India. (1876.) By RASH BEHARI GHOSE. Royal 8vo., cloth. Rs. 10.

The Law relating to Minors in Bengal. (1877.) By E. J. TREVELYAN. Royal 8vo., cloth. Rs. 10.

The Hindu Law of Marriage and Stridhana. (1878.) By GOOROO DOSS BANERJEE. Royal 8vo., cloth. Rs. 10.

The Law relating to the Hindu Widow. By TRAILOKYANATH MITTRA, M.A., D.L. Royal 8vo., cloth. Rs. 10. (1879.)

The Principles of the Hindu Law of Inheritance. By RAJCOOMAR SARVADHICARI, B.L. Rs. 16. (1880.)

The Law of Trusts in British India. By W. F. AGNEW, Esq. Rs. 12. (1881.)

CALCUTTA.—THACKER, SPINK & CO.

Corresponding Houses.

BOMBAY.—THACKER & CO. (LIMITED).

LONDON.—W. THACKER & CO., 87, NEWGATE STREET.

One Vol. Small 8vo. Rs. 3-8.

TALES FROM INDIAN HISTORY:

BEING THE

Annals of India retold in Narratives.

BY J. TALBOYS WHEELER.

Forms a complete History of India from the earliest period to the present day, drawn up as a series of "Narratives' for general reading in schools and families. So far it will resemble the history of Scotland as told in Sir Walter Scott's "Tales of a Grandfather," omitting all details and disquisitions which are sufficiently given in the author's larger histories of India, and seeking to tell the progress of events in the most simple and attractive manner.

Reviews of Wheeler's Tales from Indian History.

"While the work has been written for them (natives), it has also been written for the people of England, who will find in the volume, perhaps for the first time, the history of our great dependency made extremely attractive reading. Mr. Wheeler's narrative is written in a most graceful style: indeed, he is master of the English language. He does not confine himself to the mere dry details of history, but tells the adventures of Indian heroes and heroines in legends of love and war; describes the village communities of India, their organization and self-government; delineates the results of caste, infant marriage, and other Hindoo institutions and usages as seen in the family and social life of the people in villages and towns, as well as in courts and palaces. . . . The work also contains valuable observations on the foreign relations of the Indian Empire with Persia, Russia, Turkey, and China. Altogether this is a work of rare merit."—*Broad Arrow.*

"In going through an interesting book, the reader will be furnished with a good general notion of Indian history, and learn besides something about Indian modes of life."—*Queen.*

"Will absorb the attention of all who delight in thrilling records of adventure and daring. It is no mere compilation, but an earnest and brightly-written book."—*Daily Chronicle.*

"This little volume contains a history of India in the form of tales and narratives, intended by the author for the people of India as well as for those of the British Isles."—*Army and Navy Gazette.*

"No young reader who revolts at the ordinary history presented to him in his school books will hesitate to take up this. No one can read a volume such as this without being deeply interested."—*Scotsman.*

UNIFORM SERIES OF ILLUSTRATED WORKS.

Square 16mo. Handsomely bound. Rs. 7.

Lays of Ind. By ALIPH CHEEM. Comic, Satirical, and Descriptive Poems, illustrative of Anglo-Indian life. Seventh Edition. With 50 Illustrations by the Author, R. A. Sterndale, Lionel Inglis, and H. Berkley.

"There is no mistaking the humour; and at times, indeed, the fun is both 'fast and furious.'"—*Liverpool Mercury.*

"The 'Lays' are not only Anglo-Indian in origin, but out-and-out Anglo-Indian in subject and colour To one who knows something of life at an Indian 'station' they will be especially amusing. Their exuberant fun at the same time may well attract the attention of the ill-defined individual known as 'the general reader.'"—*Scotsman.*

"Satire of the most amu ing and inoffensive kind, humour the most genuine, and pathos the most touching pervade these 'Lays of Ind.' . . From Indian friends we have heard of the popularity these 'Lays' have obtained in the land where they were written, and we predict for them a popularity equally great at home."—*Monthly Homœopathic Review.*

Hindu Mythology: Vedic and Puranic. By the Rev. W. J. WILKINS, Calcutta Illustrated by 96 engravings from drawings by native artists and others.

"Has produced a popular guide, forming on a small scale a useful classical dictionary of India. A truthful account of things commonly believed by millions of our Indian fellow-subjects cannot be without interest, and is indeed an indispensable guide to those who would understand the history, life, and institutions of Hindustan."—*Daily News.*

"The author has attempted a work of no little ambition, and has succeeded in his attempt, the volume being one of great interest and usefulness."—*Home News.*

"No one ought to go out as a missionary nor as a civil servant without having mastered all that this book can teach him."—*The Graphic.*

"A work which everyone who takes an intelligent interest in India ought to have on his shelf." - *Times of India.*

Riding: on the Flat and Across Country. A Guide to Practical Horsemanship. By Captain M. H. HAYES, author of "Training and Horse Management in India," "Veterinary Notes for Horse Owners." With 75 illustrations.

"The book is one that no man who has ever sat in a saddle can fail to read with interest."- *Illustrated Sporting and Dramatic News.*

"Three fourths of those who indulge in what they call riding in Hyde Park would do well to learn this book by heart."- -*Vanity Fair.*

"The chapter devoted to 'ladies' riding' calls for notice in these columns, as in it will be found information of a kind which is frequently asked for, while it is not always easy to point out to the enquirer a trustworthy source from which it may be obtained. So far, however, as written directions may suffice to supplement or impress practical teaching, all that is requisite will be found in Captain Hayes' pages." —*The Queen.*

"There is left nothing unsaid in the present publication to ensure a thorough acquaintance with the subject"—*Bell's Life.*

The Ferns of British India, Ceylon, and Malaya. By Colonel R. H. BEDDOME, author of "The Ferns of British India," "The Ferns of Southern India." With 300 illustrations. (Double volume.)

"This work has been specially written in a style as free from technicalities as possible to meet the wants of non-scientific readers, and to assist visitors to the numerous Hill Stations of India in the selection and gathering of Ferns. The work will be illustrated from the drawings already so well known in India."

Other Valuable Works are in Preparation.

W. I. RICHARDSON, PRINTER, 4 AND 5, GREAT QUEEN STREET, LONDON, W.C.